Mekki ZEROUALI

Use of solar energy for drinking water production in southern Algeria

Mekki ZEROUALI

Use of solar energy for drinking water production in southern Algeria

ScienciaScripts

Imprint
Any brand names and product names mentioned in this book are subject to trademark, brand or patent protection and are trademarks or registered trademarks of their respective holders. The use of brand names, product names, common names, trade names, product descriptions etc. even without a particular marking in this work is in no way to be construed to mean that such names may be regarded as unrestricted in respect of trademark and brand protection legislation and could thus be used by anyone.

Cover image: www.ingimage.com

This book is a translation from the original published under ISBN 978-620-8-82581-2.

Publisher:
Sciencia Scripts
is a trademark of
Dodo Books Indian Ocean Ltd. and OmniScriptum S.R.L publishing group

120 High Road, East Finchley, London, N2 9ED, United Kingdom
Str. Armeneasca 28/1, office 1, Chisinau MD-2012, Republic of Moldova, Europe
Managing Directors: Ieva Konstantinova, Victoria Ursu
info@omniscriptum.com

Printed at: see last page
ISBN: 978-620-8-99186-9

Contents

Biography of the author

Mekki Zerouali holds a doctorate in energy physics from the University of Batna 1 (El-Hadj Lakhdar Batna, Algeria). He is currently assistant professor at the École nationale supérieure des énergies renouvelables, de l'environnement et du développement durable. His research focuses on the integration of solar energy in the residential building sector. An active member of the Applied Energy Physics Laboratory, he works on developing sustainable solutions to improve the energy performance of buildings. Dr Zerouali is the author of numerous scientific publications in national and international journals.

Dedicaces

Je JeJie e travatid :

My mother;

Man pèr ;

My brothers;

Mess$urs;

aus m s friends tcalègues;

Taus ceux qui m sant ciers ;

rouali Meiii.

GENERAL INTRODUCTION

Water is an essential element for the continuity and development of humanity. But what does the future hold for this important resource?

Worldwide, the issue of water supply is becoming more worrying by the day. The unanimous conclusion is simple: the situation is already precarious in some parts of the world, and can only get worse in the years to come.

The tremendous demographic growth that our planet will experience over the next twenty-five years will necessarily be accompanied by an explosion in water consumption and a deterioration in its quality and quantity. This could seriously jeopardise the supply of fresh water to a large part of humanity and, as a result, aggravate conflicts between neighbouring countries with shared resources. To this end, and to better safeguard the future in terms of releasing and mobilising water resources, it would be wiser to rethink the planning of conventional water resources as part of a global vision that also incorporates the use of non-conventional water resources, in particular the desalination of seawater and brackish water.

The desalination of brackish water and, a fortiori, seawater provides an attractive solution to this problem. That's why the desalination market is already booming, growing by 7% a year. The desalination processes that have proved their reliability can be divided into two main families: one based on liquid-vapour phase change and the other using membrane separation.

The distiller's output depends on several parameters, the most important of which is solar radiation. In order to optimise the influence of the various parameters on the performance of the double-slope solar distiller, we carried out a comparative numerical and experimental study on the variation of the parameters characterising the operation of the conventional solar distiller.

Our work is divided into four chapters. **Chapter 1** presents the situation regarding the consumption and distribution of drinking water in the world. In the second part of this chapter we present the distribution of water resources and the problems threatening the existence of drinking water in Algeria.

In the 2nd chapter, we present the different desalination technologies, i.e. the so-called electric or phase-change methods. The conventional solar distiller is one of the phase-change methods based on the principle of the greenhouse effect. It consists of a basin covered with an absorbent black layer and a transparent cover made of ordinary glass or sometimes sloped plastic. This is followed by a presentation of the various works carried out in the field of solar desalination, both numerical and experimental.

The **3rdchapter** presents the solar distiller to be studied, the simplifying assumptions, the energy balances for each part of the solar distiller, and the expressions for the heat transfer coefficients. The chapter ends with a flow chart of the various calculation steps.

The presentation, discussion and interpretation of the results are the subject of the **4th chapter.**

The work is concluded with a general conclusion summarising the various results and proposing recommendations for future work.

Chapter 1

The World Water Situation

1-Introduction

Water has been the essential element for all life on earth since the creation of the universe, so much so that it is very abundant on our planet. 97.2% of the quantity available on earth is made up of seawater that cannot be used directly. **[1].**

-Climate change and population growth in certain regions of the world are leading to a chronic shortage of drinking water.

-Available freshwater (lakes, rivers, groundwater) represents only 0.07% of the total resource, i.e. around one million km^3. But the distribution of this water is highly unequal. Ten countries share 60% of freshwater reserves, while twenty-nine others, mainly in Africa and the Middle East, are facing chronic shortages of freshwater. The main water resources distributed worldwide are shown in the following table(I)

Resources	Volume (m3)	Total per cent of water	Percentage of fresh water
Atmospheric water	12900	0.001	0.01
Glaciers	24064000	1.72	68.7
Ground ice	300000	0.021	0.86
Rivers	2120	0.0002	0.006
Lakes	176400	0.013	0.026
Marshes .	11470	0.0008	0.03
soil moisture	16500	0.0012	0.05
A(quifèr) es	10530000	0.75	30.1
Lithosphere	23400000	1.68	
Oceans	1338000000	95.81	

Table 1-1 - The main water resources distributed in the world. [1] - The countries of the Middle East and the southern Mediterranean basin are among those regions of the world where drinking water resources are extremely limited. A large proportion of water reserves are brackish. Rainfall is highly variable, both in space and time. Droughts can last for months, and are sometimes followed by torrential rains that sweep away everything in their path, such as the **Bab-Elouad** flood in Algiers in October 2001, when a thousand people were buried by a torrent of mud.

-In the following section we will briefly present statistical data on drinking water resources and the consumption of each inhabitant for each year in the world, followed by the water situation in Algeria.

2-Water in the world

3 It is estimated that the world's drinking water resources amount to 40,000 billion m for a population of 6 billion people, i.e. 6,800 m^3/inhabitant/year, although availability varies from one country to another.

For example, it is 360000 m^3/capita/year for Congo, 630 m^3/capita/year for Algeria and only 23 m^3/capita/year for Djibouti.

The scarcity of water in a country is assessed in relation to thresholds that are estimated in the water literature at 1000m^3/inhabitant/year (water stress threshold) and 500m^3/inhabitant/year (water scarcity threshold).

In 1990, twenty (20) countries, including Algeria (table 2.), were classified below the stress threshold. In fact, these thresholds are only indicative and are based on the water requirements of countries located in arid or semi-arid zones where agricultural production depends wholly or largely on irrigated agriculture [2].

country	Water availability (m3/capita)	
	1990	2025
Algeria	720	430
Morocco	1117	590
Tunisia	540	324
South τι_- Libya	1017	359
Egypt	1123	630
Syria	2087	732
Turkey	3626	2186
Spain	2849	2733
North France	3262	3044
Italy	3243	3325
Greece	5826	5840

Table-I.2: Availability of drinking water in the Euro-Mediterranean region [3].

2-1-Geographical distribution

- Water may be plentiful, but it is not evenly distributed across the globe. Nine countries share 60% of the world's annual flow. Water is also irregularly distributed from one year to the next or from one season to the next (more than 60% of the world's annual flow is generated by floods followed by droughts, sometimes lasting several years). In addition to these seasonal variations, the world's resources are characterised by significant geographical variation:
- In South America, there is a sharp contrast between the area covered by the Amazon, which drains 15% of the world's freshwater resources, and the north-east of Brazil, which suffers from drought.
- In India, there are huge differences between the plains of the Himalayas (crossed by large rivers), the desert areas and the monsoons of southern India.
- In China, the water is frozen for several months of the year in the north, the west is characterised by desertification and the south by a tropical climate.

In the Mediterranean region, freshwater is scarce and unevenly distributed. The most water-rich countries (France, Turkey, Italy, former Yugoslavia) account for two-thirds of resources (825 out of 1189 km^3per year). Other countries, such as Malta, Jordan and Libya, are below the scarcity threshold (500 m^3per year per inhabitant).

2-2-Water consumption

There are many uses for water

Agriculture accounts for around 70% of all freshwater consumption on the planet. Most of this consumption is accounted for by irrigated agriculture, which occupies around 17% of cultivated land and accounts for 40% of the world's agricultural production (the remainder is provided by rain-fed agriculture). Irrigated areas have almost doubled worldwide since 1960.

- Industry is responsible for around 20% of the world's freshwater consumption. It takes 80 litres of water to produce 1 kg of steel, 1250 litres to produce 1 kg of aluminium and around 8,600 litres to produce a six-inch memory card.

- Domestic consumption accounts for 8-10% of total freshwater consumption.

The ways in which water is used do not all have the same consequences. The use of natural resources is considered less destructive when the water is available again after use, as in the case of domestic water that is treated and returned to watercourses. On the other hand, evaporation and infiltration prevent water from being reused immediately.

2-3-WHO standards for drinking water.

Drinking water must meet very strict quality standards based on the medical work of the World Health Organisation (WHO) **[5]**. These criteria can be grouped into six categories:

- Organoleptic parameters: transparency, colour, odour and flavour.
- Physico-chemical parameters :

-PH, temperature, mineral inorganic concentrations, conductivity.

-Undesirable substances (nitrates, fluoride, etc.).

-Toxic substances (lead, chromium, etc.).

- Microbiological parameters: harmful bacteria (coliforms, faecal streptococci, etc.)
- Pesticides and related products.
- Parameters concerning softened water supplied for human consumption.
- Parameters relating to undesirable substances, substances tolerated in very small quantities that can have an impact on health without causing any short-term inconvenience.
- These standards are grouped together in two tables

Parameters	**European standards** (mg/l)
Aluminium ($Al^{(+3)}$))	max 0.2
Chlorides	max 200
Magnesium (Mg+2)	max 50
PH	from 6.5 to 9
Potassium (K)$^{+}$	max 12
Sodium (Na)$^{+}$	max 150
Sulphates ($_{SO-}$)	max 250

Table -I.3: Physico-chemical parameters of drinking water [5].

Toxic substances

Parameters	**European standards**
Arsenic As	max 50 Lig/l
Cadmium $_{Cd2+}$	max 5gg/l
Soluble chromium Cr	max 50ug/l
Cyanides Cn^{-}	max 0.05 μβ/l
Mercury Hg	max 1 μβ/l
Nickel Ni	max 50 μβ/l
Phosphorus P	max 5 μβ/l
Lead Pb	max 50 μβ/l

Table -I.4: Substances toxic to water [5].

2-4-Problems threatening water. [6]

2-4-1-Pollution

Pollution is one of the major problems threatening water. By definition, water pollution is any chemical, physical or biological change in water quality that has a harmful effect on living organisms. Water pollution can be divided into three categories **[6]**:

- The first concerns disease-causing agents. These include bacteria, viruses, protozoa and parasitic worms that develop in sewers and untreated wastewater.
- The second is waste that can be decomposed by bacteria that need oxygen. When these bacteria are in action in large numbers, they can lower the oxygen level in the water, leading to the death of other species living in the water, such as fish.
- The third is water-soluble inorganic pollutants, such as acids, salts and toxic metals. Large quantities of this type of compound make the water unsuitable for consumption and cause the death of aquatic life.

2-4-2-Climate change

The quantity of gas (CO_2, C'[[4,....) is increasing in the atmosphere (due to the use of fossil fuels), trapping solar radiation (heat) emitted by the earth, which increases the earth's temperature - the greenhouse effect. The first signs of global warming are visible: floods, desertification, the spread of disease and the disappearance of animal species. If nothing is done, scientists predict that the concentration of CO_2 will double over the course of the 21st century, making climate change even worse. This is the chronicle of a disaster foretold.

2-4-3-Waste

On average, an African wastes 5% of the quantity consumed, a consequence of poor management of water resources at a time when public spending in all African countries is struggling to reach 0.5% in the water resources sector.

2-4-4-Resource depletion

At present, some 5,000 people around the world die because they have no access to water. More fears relate to the period around 2050. By then, the world's population is expected to reach 10 billion, and supply problems are likely to deprive half of this population of adequate water resources. One technical response is to reduce water consumption by spreading good practices and using technologies that consume less water. But the main problem lies in the distribution of drinking water and the consequences of its absence in certain areas.

In many countries, annual groundwater abstraction already exceeds aquifer recharge:

- In the United States, the Ogallala aquifer, which stretches from South Dakota to Texas, covers an area the size of France and is one of the world's largest underground water reserves, is being drained 8 times faster than it is being filled because of the 200,000 wells that tap into it to irrigate 3 million hectares of crops.
- In Mexico City, the water pumped to supply the city exceeds the capacity of the water table by more than 50%.
- Since the 1960s, the Aral Sea has lost 60% of its surface area as a result of the USSR's efforts to develop cotton growing in Central Asia (particularly in Uzbekistan) (at the time, it was considered to be the 4th largest closed body of fresh water in the world).
- Since the early 2000s, the level of the Dead Sea has been falling by one metre a year as a result of withdrawals from the Jordan River.
- In Africa, Lake Chad has lost 80% of its surface area.
- The draining of the Yellow River, which drains the immense cereal-growing basin of

northern China, means that the lower reaches of the river dry up for more than 200 days in some years.

3-Drinking water in Algeria

-Algeria is one of the poorest countries in terms of water potential, and is therefore below the theoretical scarcity threshold set by the World Bank, i.e. 1000 m^3per capita per year. In 1962, annual per capita water availability was 1,500 m^3, but by 1990 it had fallen to 720 m^3, 680 m($^{3)}$in 1995 and 630 m^3in 1998.annual per capita water availability is now 500 m^3. As a result of demographic pressure, this availability will only be 430 m^3per inhabitant in 2020; by this date, availability per river basin will be as follows (table -3).

River basin	Resources (hm)	Population (10^6inhabitant)	Availability (m^3/inhabitant)
Oranie chott Chergui	1400	6.3	220
Chélifer	2072	7.0	300
Algérois Soumma-Hadna	5125	15.8	320
Const-mejd Mellegue	5048	10.0	500
South	5436	4.9	1120
Total Algeria	64518	44.0	430

Table-I .5: Per capita water availability in Algeria in 2020

4-The water problem in Algeria

-This is a perception that is not yet fully understood, and is due to a number of factors, essentially

4-1-lack of water resources

4-1-1-Surface water resources

-Water resources are generally scarce and, above all, extremely irregular, and exploiting them is becoming increasingly difficult in our country. They are unevenly distributed across the country, and are also exposed to ever-increasing pollution risks, which compromise the use of these waters in many parts of the country. There are only 49 dams in operation throughout the country, with a total capacity of 4.48 billion m^3, for supplying drinking water.

It should be pointed out that the three dams supplying the capital, namely the Keddara, Hamiz and Beni-Amrane dams, have entered what is known as a "dead period", with a stored volume at the end of January 2002 of around 8.8 billion m^3, a drop never experienced since the dams were commissioned. The same situation applied to the Hamiz dams, with a volume of 1.91 million m^3, and the Bni-amran dam, which is a transfer dam to Kaddara, with a volume of 1.18 million m^3. The dams in the centre of the country are the ones most affected by the significant drop in water levels, unlike those in the west.

4-1-2-Underground water resources

-Groundwater is generally scarce and localised in the Sahara. The wadi beds feed a few water tables, often brackish: Ghir, M'zeb, Saoura, or deep, semi-fossil or fossil water tables such as the Albian. In the north, rainfall is irregular and poorly distributed geographically, increasing from west to east and successively affecting the Tlemcen mountains, the northern slopes of the Ouarsenis, the Grand Kabyle massif, Little Kabylie, northern Constantinois and the Aurès massif [7].

4-2-Continuous deterioration of resources

-The degradation of water resources is beginning to reach worrying proportions in the Tellian region, where most of the surface water potential is located. The pollution of certain dams, both by domestic waste water and by industrial discharges, has already been highlighted (Table-3-).

Dam	Wilaya	Source of pollution
Cheffia	El-Taref	Urban and industrial waste
Zardeza	Skikda	Urban waste
Hammam-Grouz	Mila	ENAD* Chelghoum-Laid
Bèni Amrane	Boumerdes	ENAD*Lakhdaria/Hydrocarbons
Kaddara	Boumerdes	Hydrocarbons
Hamiz	Algiers	Hydrocarbons
Harbil	Médéa	SAIDAL** complex
Fergoug	Mascara	Urban and industrial waste Urban
Beni-Bahdel	Tlemcen	and industrial waste Industrial
SidiAbdelli	Tlemcen	waste
(*)Entreprise National des Détergents et Produits d'Entretien. (**) National Pharmaceutical Production Company.		

Table -I.6: Dams affected by pollution in Algeria.

If nothing is done to remedy the situation, pollution is likely to be one of the serious causes of water shortages in the future. [8]

4-3-Drought periods

Algeria has been experiencing severe and persistent drought for over twenty years. 1988/89 was a dry year. The rainfall deficit was particularly marked in the west (especially in Oran, Ghazaouet and Arzew).

- The Beni-Bahdel and Mafrouch dams have dried up. During the month of June 1988, the Central and Western regions have seen a rainfall deficit of over 50%, while in the East it has been 30% for the last two decades. Generally speaking, drought has affected most countries in the Mediterranean basin since the early 1980s.

4-4-Hard weather conditions

- Algeria's climate is semi-arid (200mm to 500mm), which means that resources are increasingly limited and difficult to exploit. The national territory is characterised by three highly contrasting climatic zones:
- The coastline and mountain ranges account for around 4% of the total surface area. The climate is Mediterranean, with very heavy rainfall in winter causing severe erosion. In summer, rainfall is extremely rare and the heat is very intense. Rainfall can reach 1600 mm/year on the higher ground, but is irregular from one year to the next and unevenly distributed.
- The high plateaux, which occupy around 9% of the total surface area, are characterised by a semi-arid climate (rainfall of between 100 and 400 mm/year), resulting in limited water resources.
- The Sahara, an arid desert with an average rainfall of less than 100 mm/year, covers 87% of the country, with extreme climatic conditions and very wide temperature ranges. **[9]**

4-5-Population growth

Algeria had around 32.3 million inhabitants in 2004, 41% of whom live in rural areas. The average density is 14 inhabitants/km^2, but the population is highly concentrated in the coastal zone, which is made up of more fertile agricultural land rich in natural resources, as well as in the country's main towns and centres of economic activity (215 inhabitants/$km^{2)}$, compared with 38 inhabitants/km^2in the high plateaux and 7 inhabitants/km^2in the south. Today, the population exceeds 35 million.

4-6-Losses in irrigation networks

-Total losses in irrigation networks are estimated globally at 40% of abstractions; in towns they are estimated to exceed 50% and generally fluctuate between 30% and 70%. This phenomenon, so widespread as to have become commonplace, has ended up giving the impression that it is a natural inevitability. We need to learn not to waste water on the one hand, and to produce safe water from seawater or heavily polluted sources on the other[8].

4-7-Rate of connections to the drinking water network

-The rate of connection to the drinking water supply is 85%. The cost price of a cubic metre of water is 20DA, while it is sold at 3.80DA for the 1sttranche (<25 $m^{3)}$, and 24.70DA for the 4thtranche (>82 m3). Water losses are estimated at 50%. Distributed water consumption is 161 l/capita/d; if leaks (50%), industry and tourism are taken into account, this quantity becomes 60 l/capita/d, (Table 4). **[10]**

The year	1962	1990	1995	1998	2000	2020
m^3/inhab.	1500	720	680	630	500	430

Table -I.7: Availability of drinking water in Algeria

5- Different types of water

5-1- Drinking water

According to WHO (World Health Organisation) international standards, water intended for domestic use must have a salinity of less than or equal to 500 mg/l. It must also contain no more than 250 mg/l of chlorides and no more than 250 mg/l of sulphates.

For industrial and agricultural uses, standards vary widely, but in most cases it is desirable for salinity not to exceed 1000 to 1500 mg/l.

5-2- Sea water

Seawater is a liquid with an extraordinarily high composition, containing around fifty simple elements, ranging from hydrogen, to oxygen, chlorine and sodium. Generally speaking, seawater contains around 35 g/l of dissolved mineral salts, although the salinity may be different in the case of enclosed seas:

- Mediterranean Sea: 36 to 39g/l,
- Red Sea: around 40g/l,
- Caspian Sea: 13g/l,
- Dead Sea: 270g/l,
- Persian Gulf: 36 to 39g/l

Note

- The average PH of seawater varies between 7.5 and 8.4, making seawater a largely basic medium.

5-3 Brackish water

Generally speaking, brackish water is non-drinkable saline water with a salinity lower than that of seawater. Most brackish water has a salinity of between 1 and 10 g/l. It occurs either as surface water or groundwater.

5-4 -Distilled water

Distilled water is a pure substance. It is generally used in :

- Laboratories and hospitals for all analyses, vaccines, serums, etc.
- Certain industries for use in battery factories; processing requiring the use of pure water (photographic industry)

5-4-1-Purity criteria for distilled water

These are physical constants with well-defined values that characterise distilled water. The density of distilled water is always equal to 1g/cm^{3}or 1000 kg/m^{3}at a temperature of 4°C. Distilled water gives absolutely identical pure water.

Pure water is a colourless, odourless liquid. It is not drinkable because it is tasteless; to taste pleasant, water must contain 0.1 to 0.5g per litre of dissolved minerals and an oxygenation level of 6.4mg/l.

5-2- The thermo-physical properties of water

-Thermal conductivity

Its expression is given by the following relationship

$$\lambda(T) = 0.569 + 0.185.10^{-2}.T + 10^{-5} * 0.749.T^2 \quad \text{(T en degré K}^0\text{)} \quad (1.1)$$

Latent heat of vaporisation

I lie is defined by the expression :

$$L_v(T) = 3408 - 5.21T + 0.01 + T^2 - 1.194T^2) \quad \text{(T en degré K}^0\text{)} \quad (1.2)$$

Niassic heat at constant pressure

It is given by :

$$C_p(T) = 4216.66 - 3.165.T + 0.886.10^{-2}.T^2 - 0.101.10^{-2}.T^3 + 0.445.10^{-5}.T^5 \quad (1.3)$$

Dynamic viscosity coefficient :

$$m(T) = 5.449 - 0.167\,T + 7.325.10^{-4}T^2 \quad (1.4)$$

The ma%e by volume :

$$\rho(T) = 999.879 + 0.0469.T - 0.667.T^2 + 0.208.10^{-4}.T^3 \quad (1.5)$$

Coefficient of kinematic viscosity :

$$n(T) = \frac{m(T)}{\rho(T)} \quad (1.6)$$

$$n(T) = 1.7414.10^{-6} - 4.453.10^{-8}.T + 5.412.10^{-10}.T^2 - 3.3955.10^{-12}.T^3 \quad (1.7)$$

The coefficient of thermal expansion

$$\beta(T) = (-0.884 + 0.118.T - 0.26.10^{-2}T^2 + 0.252.10^{-4}T^3 - 0.975.10^{-7}T^4).10^{-4} \quad (1.8)$$

Chapter 2

The different desalination technologies... 1-Introduction :

-Faced with a definite water crisis that is beginning to make itself felt throughout the world, in addition to the economic constraints on sustainable development, appropriate solutions need to be devised to prepare for this challenge that threatens the very existence of mankind. Our country, which has considerable saline water resources and an equally large solar deposit, needs to use desalination techniques, the reliability of which no longer needs to be demonstrated, combined with renewable energy sources. This solution is a fairly reliable way of producing drinking water. However, it should be pointed out that this method remains economically reliable only for small capacity units, ranging from a few $m^3$to tens (or even hundreds) of $m^{(3)}$ ofdrinking water per day.

-For large capacities, combining desalination processes with renewable energies requires high investment costs, and the reliability of such systems is not always guaranteed.

2-The main water desalination technologies

-Desalination processes fall into two main categories: on the one hand, membrane processes or electrical methods (filtration). On the other hand, distillation or thermal methods (which require a change of phase, evaporation/condensation). Desalination plants comprise four stages, regardless of the desalination process used:

1-A seawater intake via a pump and coarse filtration.

2-A pre-treatment with finer filtration, the addition of biocides and anti-tartar products. [11].

3- The desalination process itself.

4-Post-treatment, including remineralisation of the pure water produced. At the end of these 4 stages, the seawater is made drinkable or industrially usable, and must contain less than 0.5 g of salts per litre.

Before mentioning the different desalination methods, the following figure classifies the desalination processes according to the quantity of water treated and its salt content.

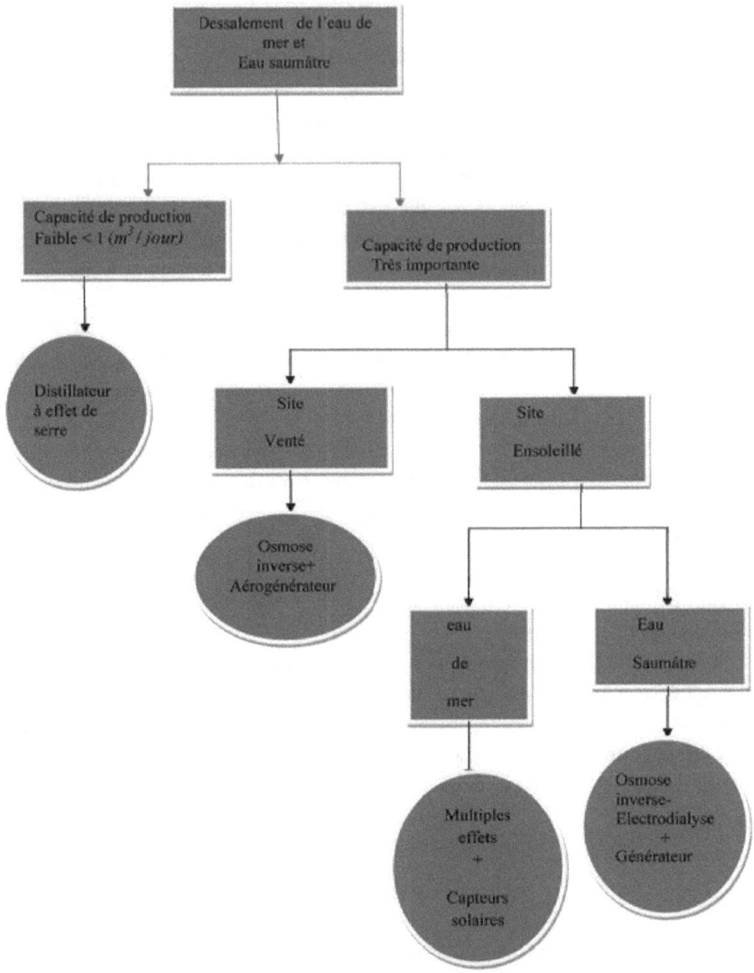

Figure -2-1 : Desalination processes. [12]

2-1- Processes using membranes

There are two categories:

.2-1-1-Reverse osmosis

This is a process for separating water from dissolved salts by means of semi-permeable membranes under the action of pressure (54 to 80 bars for seawater treatment). This process operates at ambient temperature and does not involve any phase change. The polymer membranes used allow water molecules to pass through and prevent the passage of dissolved salts and organic molecules up to $10^{(-7)}$ mm in size. The phenomenon of osmosis will result in a flow of water directed from the dilute solution towards the concentrated solution. If we try to prevent this flow of water by applying pressure to the concentrated solution, the quantity of

water transferred by osmosis will decrease. At some point, the pressure applied will be such that the flow of water will be cancelled out. If, for simplicity's sake, we assume that the diluted solution is pure water, this equilibrium pressure is called osmotic pressure.
An increase in pressure above the osmotic pressure will result in a flow of water in the opposite direction to the **osmotic** flow, i.e. from the concentrated solution to the dilute solution. This is the phenomenon of reverse osmosis. The energy required for reverse osmosis is purely electrical, consumed mainly by the high-pressure pumps. The salt content of reverse osmosis water is around 0.5 g/l.

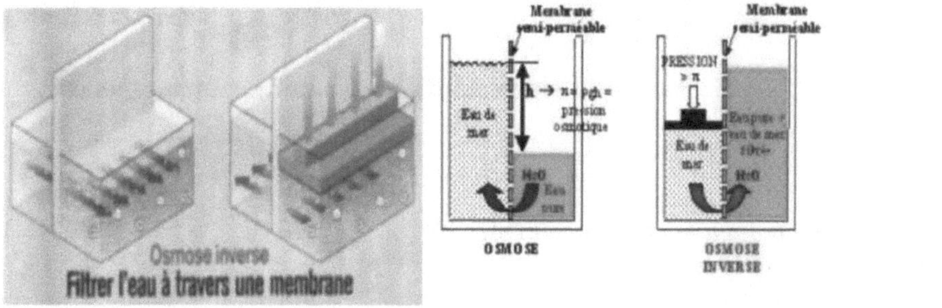

Figure II.2: Principle of reverse osmosis [13]

2-1-2-Electrodialysis

2-1-2-1-Description of the process

Ions are extracted by migrating through selective membranes (anionic or cationic) under the action of an electric field. This means that only anions can pass through an anionic membrane and only cations can pass through a cationic membrane.
cationic. By placing several membranes in parallel, allowing positive and negative ions to pass through alternately, certain ions can be removed from the water.
-Particles that do not carry an electrical charge are not eliminated.

2-1-2-2-Application to desalination

-Electrodialysis is used for seawater desalination. Figure 3 illustrates the operation of a parallel desalination unit. At the end of the treatment chain, both fresh water and brine are recovered. There are also series units and other models of electrodialysis system.

2-1-2-3-Operating principle

Under the effect of a current applied in the external tanks, Na^+ ions are attracted to the negative electrode and Cl^-ions to the positive electrode. Because of the selectivity of the membranes; Na^+ ions can only pass through the cathodic membranes and Cl^-ions through the anionic membranes, fresh water is obtained in two of the four Compartments. See figure-3-

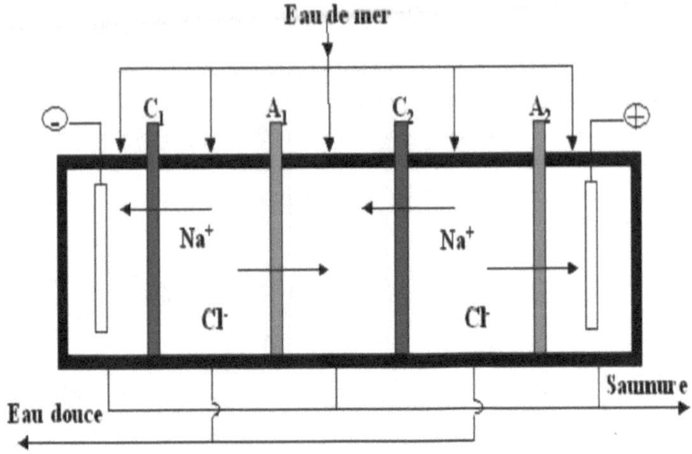

Figure II.3: Electrodialysis unit [14].

-2-2-Thermal processes

We can mention :

-Freezing

- The Multi-Stage Flash distillation (MSF) process.
- The Multi-Effect Distillation (MED) process.
- Solar distillation.

2-2-1-Freezing.

This technique is used in warm countries. It is a vacuum freezing process, the principle of which is quite simple. When salt water is frozen, the crystals that form are practically pure ice, the salt being expelled from the water during crystallisation. This forms a brine on the surface of the crystal. The crystals are then washed and melted, producing fresh water. Another method based on freezing is the secondary refrigerant method, in which heat transfer is achieved by using a liquid hydrocarbon that is immiscible with water. The mixture of butane and seawater passes through a heat exchanger where the heating causes the butane to boil, resulting in the freezing of the water in the brine. The ice crystals are washed, separated and finally melted by heat exchange with the compressed butane vapour. [15].

2-2-2- The Multi-Stage Flash distillation (MSF) process

This process consists of keeping the water under pressure throughout the heating process. When it reaches a temperature of around 120°C, it is introduced into an enclosure (or stage) where the pressure is reduced. The result is instantaneous vaporisation by expansion, known as flash. A fraction of the water evaporates (Figure 4) and then condenses on the condenser tubes placed at the top of the enclosure, while the liquid water is collected in receptacles below the tubes.

It is the hot salt water that cools down to provide the heat of vaporisation. Boiling stops when the salt water has reached the boiling temperature corresponding to the prevailing pressure. In the stage under consideration. The flash phenomenon is then reproduced in a second stage at an even lower pressure. The water is thus vaporised by successive expansion in a series of stages at increasingly reduced pressures. There can be up to 40 successive stages in an industrial MSF unit.

The main advantage of the MSF process is that the salt water does not evaporate around the heating tubes, as the liquid "flashes", which reduces the risk of scaling.

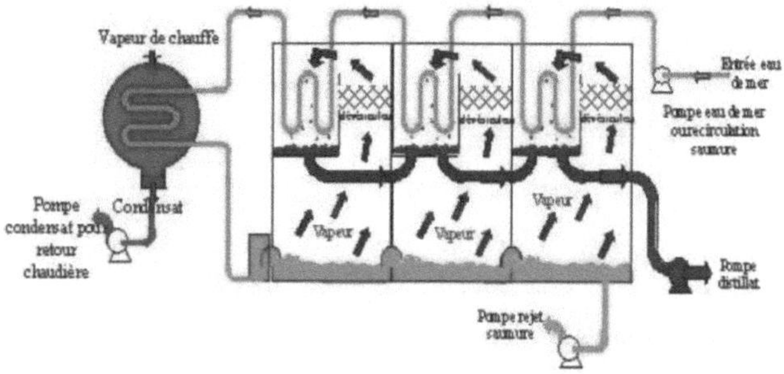

Figure II.4: Operating principle of a 3-stage (MSF) system [16].

2-2-3-The multi-effect distillation process (MED)

This process is based on the principle of evaporation, under reduced pressure, of part of the salt water preheated to a temperature of between 70 and 80°C. The water is evaporated on an exchange surface, unlike the previous process, where evaporation takes place by expansion within the successive stages. The heat transferred across this surface is provided either by steam produced by a boiler, or by hot water from a heat recovery unit.

The steam produced in this way in the 1st effect is condensed to produce fresh water in the ^{2nd}effect, where the pressure is lower. The heat of condensation it gives off makes it possible to evaporate some of the seawater contained in the 2nd effect, and so on (Figure .5). Only the energy required for evaporation in the first effect is of external origin, so increasing the number of effects reduces the specific consumption (energy/$m^{(3)}$ ^{of}fresh water produced).

Several multi-effect evaporator technologies exist:

- **2-2-3-1- Multiple-effect evaporators with sprayed horizontal tubes**

These are the devices most commonly used today. In these devices (Figure .5), the heating fluid flows through the horizontal tubes, while the salt water to be evaporated is sprayed so that it flows in as uniform a film as possible over the outside of the tubes. The vapour produced in the calender (cylindrical enclosure containing the bundle of tubes) is then sent to the tubes of the next effect, where it will give off its condensation energy. These evaporators have a very good heat exchange coefficient thanks to the film-like flow of seawater.

- This is why they are now replacing the older flooded tube bundle evaporators, in which the tubes were immersed in seawater. Figure 5.

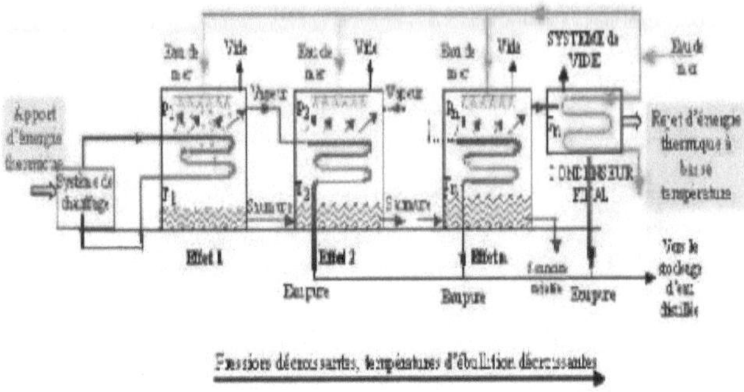

Figure II. 5: Horizontal tube multi-effect evaporator system [source1].

2-2-3-2-Multiple plate effect evaporators

They are currently being developed. In this type of device, the seawater to be evaporated flows in the form of a thin film along a thin metal plate heated by the steam from the previous effect, which flows along the other side of the metal plate. Numerous plates between which the seawater and the heating steam flow alternately are associated in parallel to form an effect. The steam produced is collected in a cylindrical calender, in which the plates are placed. This steam is then sent between the plates located in a calender, which constitutes the second effect, and so on. [17].

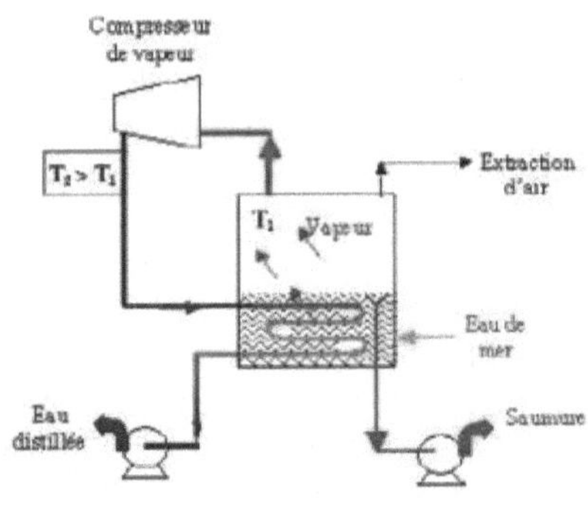

Figure II.6: Multiple-effect plate evaporator system. [Source1]

2-2-4-Vacuum blasting

The salt water in (A) (figure .7) cooled by the heat exchangers (B1, B2) enters the lower compartment of the converter (C), where the pump (d) maintains a high vacuum; it mixes with the cold brine entering through (E) and both boils and forms ice crystals. The steam escapes through (F) and the crystals are carried to the scrubber (G) where fresh water from (H) washes them out; the brine passes to the axial compartment (I) to be partly recycled and rejected after cooling the exchanger (B1).

The refrigeration unit (J) maintains the low temperature of the washer, from where the ice crystals, collected by the rotary scraper (K), pass to the converter to be melted by the heat of condensation of the steam; fresh water at 0°C, after passing through the heat exchanger (B1, B2), is collected in (L). [18]

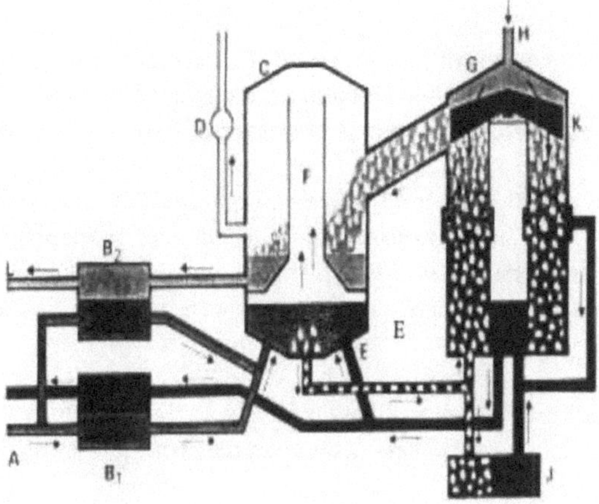

Figure II.7: Vacuum desalination system. [Source 2]

2-2-5-Desalination using solar energy

2-2-5-1-The history of solar desalination.

The idea of making pure water from seawater has plagued thirsty populations for hundreds, if not thousands, of years. The original premise was based on the idea that water could be separated from salt by boiling or evaporation. This theory - evaporation or distillation - formed the basis of the technology of the first large-scale desalination plants that appeared in the 50s and 60s, initially in the Middle East.

But the greatest achievement came in 1872 near las -salinas in the north of Chile.

-The ^{1st}module was built by Carlos Wilson. Swedish engineer with a glass surface of 5000 m^2producing up to 20 $m^{(3)}$ ^{of}drinking water per day from a saline well containing 140 (g/l).

-This desalination system remained in operation until 1990, due to the accumulation of salts in the basin, which required regular cleaning of the distiller.

Solar distillation experienced a slack period between 1880 and the First World War. In 1920 Kaush used metal reflectors to concentrate the sun's rays, resulting in the appearance of new equipment (roof type, V-shaped cover, inclined plate....) - An installation by Guinitous grouping 30 1m^2units was built at Bengardane in Tunisia in 1929. At the beginning of 1930 Trifinov proposed an inclined distiller, and in 1938 Abbot used cylindrical-parabolic reflectors to concentrate the sun's rays, the whole system being channelled through tubes containing water.

Also for Women a task in this field, Maria Tleks invented in 1945 a new type of solar distiller is called spherical solar distiller, which was used the way, American a large number of this model, about 2000 used during the 2^{em}World War .in 1969 Cooper proposed a simulation to analyze the performance of a greenhouse distiller, Since the 70s several types of solar distillers have been developed and studied, including :

The multiple-effect distiller

-The tray tilts or the distiller tilts

The loaf-type distiller or multiple loaf distiller

-Greenhouse effect distiller combinations

Several countries are interested in solar desalination, including
-Algeria
In 1953 Cyril **Gomela** developed different types of solar distiller, more than 20 solar distillers of about ten types were tested and some others were marketed throughout North Africa, the single and Australia, tests were carried out in the dessert of Biskra, Adrar and Ouargla.
-Australia
The Csiro (as wealth sientific industrial Research organization) undertook research on this type of distillation. A unit similar to the one in Gomela was developed between 1963 and 1967, the Csiro built and tested more than 8 distillers, the aim of these tests was to improve the efficiency of the distiller, a variety of materials were used for the construction of the distiller.
-India
- In 1957, two small greenhouse distillation units were built to study the effects of glass inclination, and it was found that a glass distillation unit with a low solution thickness gave good yields.
-Chile
- In addition to the first distiller built in 1982, other inclined distiller sites were installed between 1969 and 1970. The aim of this work was to predict the operating characteristics of a distiller under different atmospheric considerations.
The USA
- Research centres, including the University of California, undertook research into solar distillation just after the 2nd World War. Several configurations of simple solar distillers were built and tested in an attempt to reduce costs and increase yield.
Spain
- In 1964, Spain built the 1st desalination plant in Lanzarote, in the Canary Islands,
-In the 21^{th}century $95^0/0$ research in the field of solar desalination was carried out with the aim of improving the efficiency of desalination, and developing new techniques for producing drinking water from saline water. [19]

2-2-5-2-The solar distiller

It is a system that converts solar energy into heat energy, which is used either directly (for heating) or converted into mechanical energy via thermodynamic cycles. It consists mainly of :

- A transparent cover, generally glass, which allows solar radiation to pass through (0.4<À<0.8pm) and becomes opaque to infrared rays.
- A tank containing salt or brackish water.
- The absorber (surface coated with black paint), which absorbs as much of the sun's rays as possible, resulting in an increase in its temperature.
- Insulation to reduce heat loss.

2-2-5-3-The operating principle of the solar distiller

Solar distillation is a technique that uses the sun's rays to heat brackish water in a tank covered by sloping glass. The darker the tank, the faster the water in it heats up.) [20], and as the temperature rises some of the water evaporates and the water vapour liquefies on the inside surface of the transparent glass and is finally recovered in a recuperator (see Figure 8).

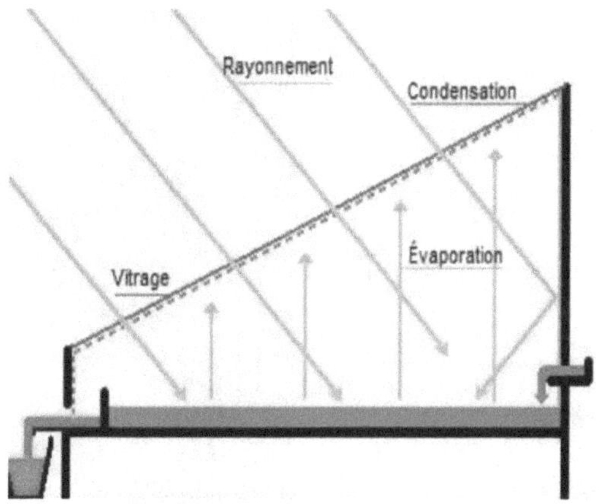

Figure II.8: The operating principle of the planar solar distiller. [Source 3].

2-2-5-4-Types of solar distillers

The different types of solar distiller can be classified according to the quantity of water produced per day or the location of the salt water.

2-2-5-4-1- First classification

Depending on production capacity, there are two types of distiller that can be built:

- Either in the form of a modular product, mainly a tank (plastic, sheet metal, wood, etc.) insulated at the bottom and covered with glass. Several distillers can be fed simultaneously to form a distillation unit. The number of distillers depends on the desired capacity of water produced. This model is only used for very small capacities (several dozen litres per day). It is practical when distilled water requirements are not very high (analysis laboratory, car park, etc.).
- There are many different types, including flat, cascade, wick, multi-effect and spherical distillers, etc.
- Or in masonry, when requirements are greater (several hundred litres/day), the surface area depending on the quantity of distilled water required. A number of applications are generally carried out in rural areas where floor space is available.

2-2-5-4-2-Deuxième classification

Greenhouse solar distillers can also be divided into two categories:

2-2-5-4-2-1-Static distillers

The water layer is immobile; they can be single- or double-glazed. These include: **2- 2-5-4-2-1-1 Single-glazed distillers**

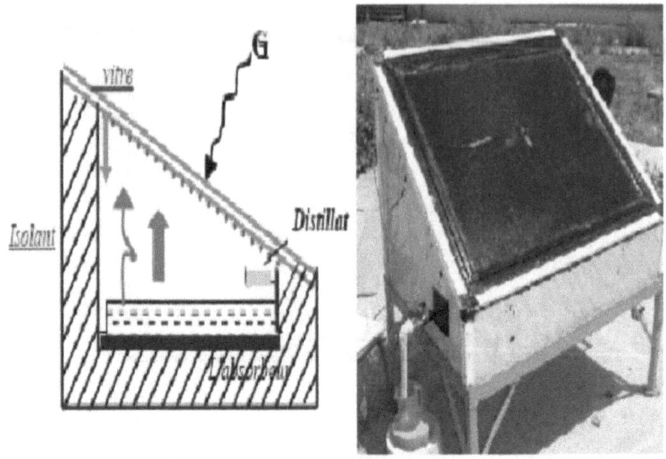

Figure II.9: Single-glazed distiller. [Source 4]

2-2-5-4-2-1-2 The double-glazed distiller

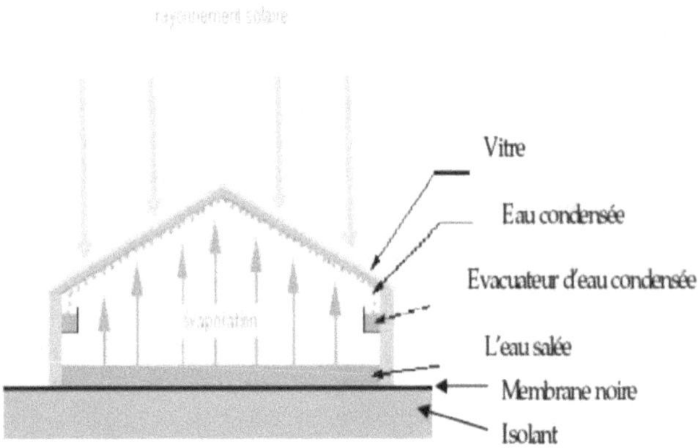

Figure II.11: Double-glazed distiller. [*Source* 6]

The two distillers have the same evaporation surface but different condensation surfaces. The advantage of the double-glazed distiller is that one collector is exposed to the sun and the other to the shade to speed up condensation. [21].

2-2-5-4-2-1-3-Multi-stage solar distiller with heat recovery Unlike the simple distiller described above, the distillation system with heat recovery consists of two different units: the heat source (a solar collector or similar) and the distillation unit. The distiller is based on the construction of a frame consisting of several basins containing salt water, stacked one on top of the other and referred to as stages (see figure). [22]

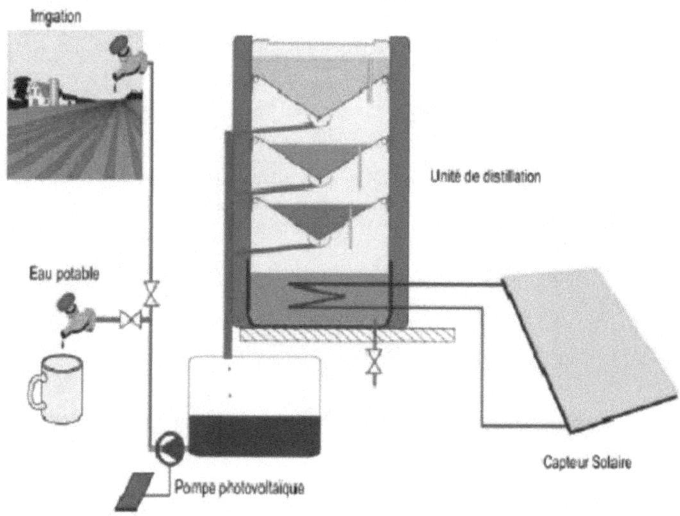

Figure I1.12: Schematic diagram of a multi-stage solar distiller. [22] **2-2-5-4-2-1-4-Cascade distiller**

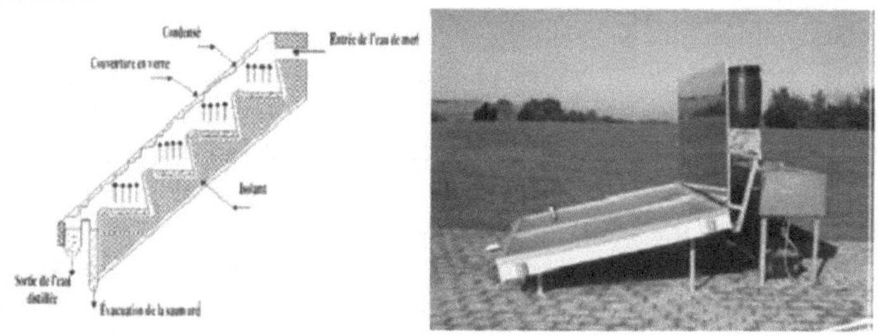

Figure II.13: Cascade distiller source. ***[Source 7]***

-This distiller has the same evaporation and condensation surfaces as the first two. It differs in the distance between condensation and evaporation. [23]

2-2-5-4-2-1-5-The spherical sweep distiller

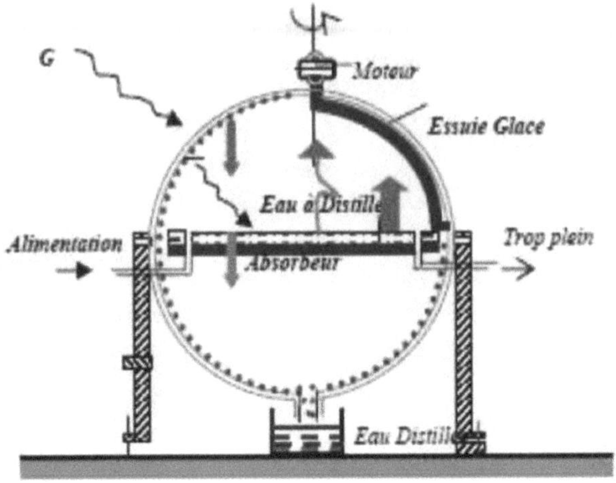

Figure II.14: The spherical sweep distiller. [24]

It consists of a sphere made of Plexiglas (or other transparent plastic material), resting on supports. Condensation takes place on the upper half-sphere, and the condensate is collected in the lower half-sphere. [25]

2-2-5-4-2-1-6-Capillary film solar distiller

It consists of a very thin layer of fine-mesh textile, soaked in water, which maintains itself in contact with the plate acting as an evaporator. The forces of inter-facial tension, which are far greater than those of gravity, play a very important role in the formation of the capillary film, which has low thermal inertia, preventing the formation of air bubbles.

This distiller consists of the following components: the collector-evaporator, the condenser and the power supply. The first cell captures solar radiation passing through the cover. The vapour condenses on the opposite wall and the heat released by this condensation allows evaporation of the film, which runs off onto the other side of the same wall. **[26]**

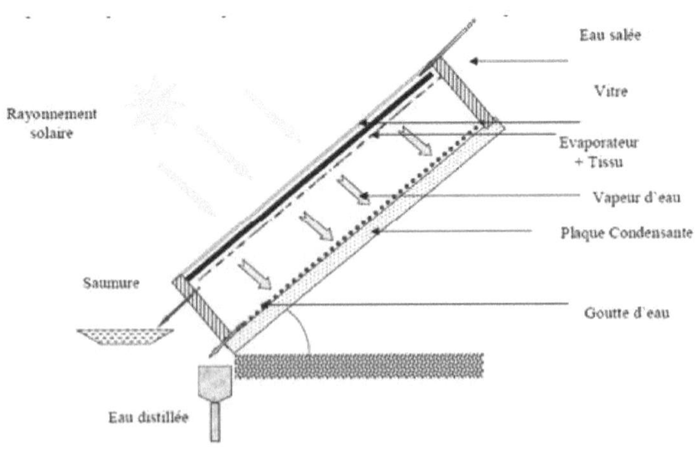

Figure II-15 Capillary film solar distiller. ***[Source 5]***

2-2-5-4-3-Trickling distillers

In this type of distiller, the water to be distilled is in motion. Different geometries exist, the most commonly used being :

2-2-5-4-3-1-Equi-current distiller

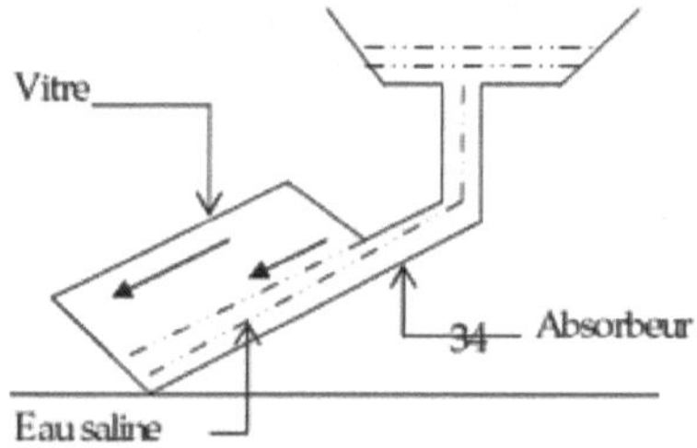

Figure II.116: Equi-current distiller.

2-2-5-4-3-2-Cross-current distiller

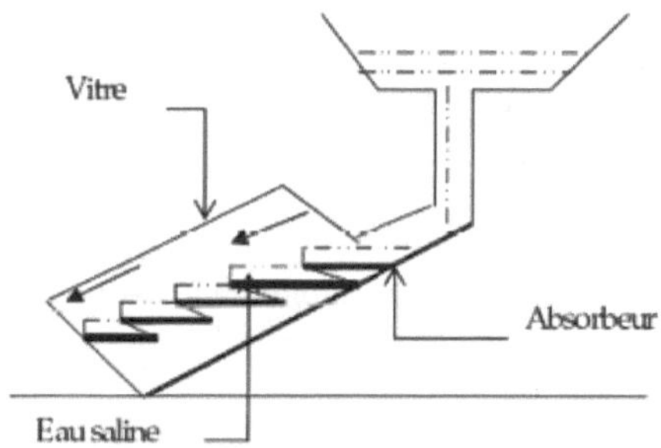

Figure II.117: Cross-flow distiller

2-2-5-4-3-3- Countercurrent distiller

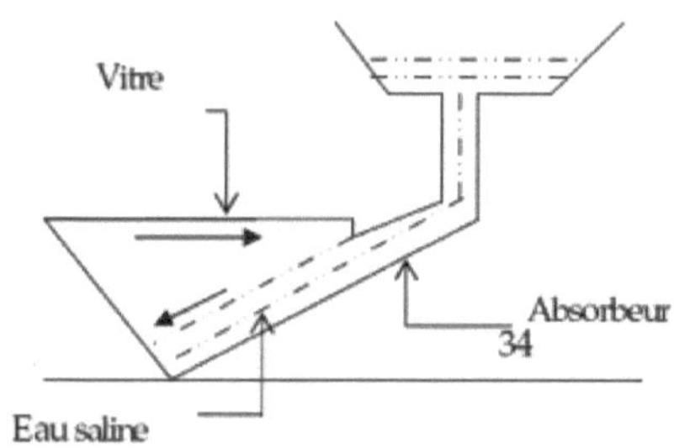

Figure II.15: Countercurrent distiller. [Source 5]

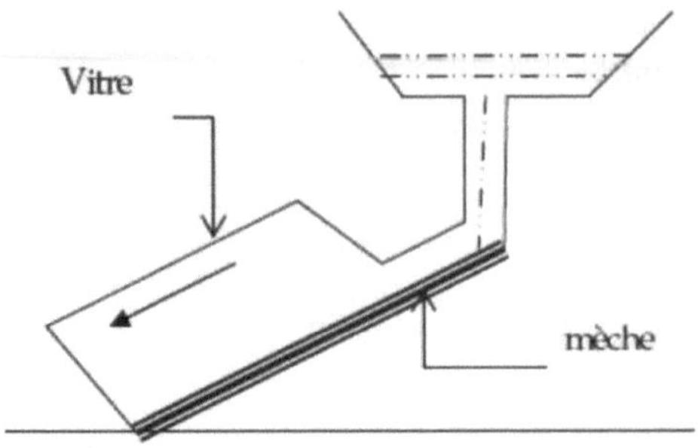

Figure 118 18: Single and single-glass wick distillers. [Source 5]

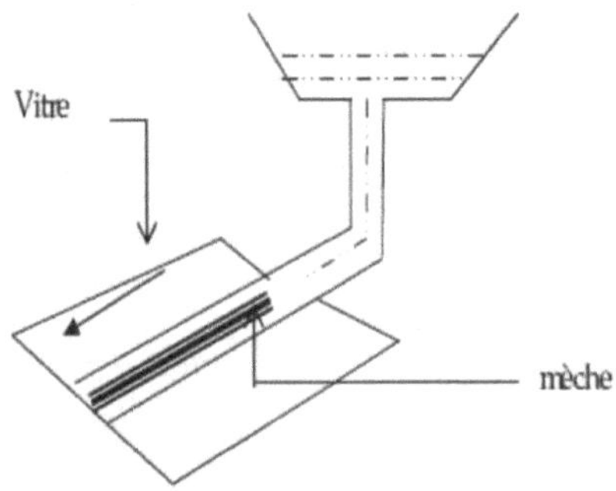

Figure 119 19: Single and double wick distillers. [source 5]

The solar distiller with double glazing and multiple wicks gives an output 20% higher than that of a distiller with a single wick.

2-2-5-4-3-5-Self-powered distiller

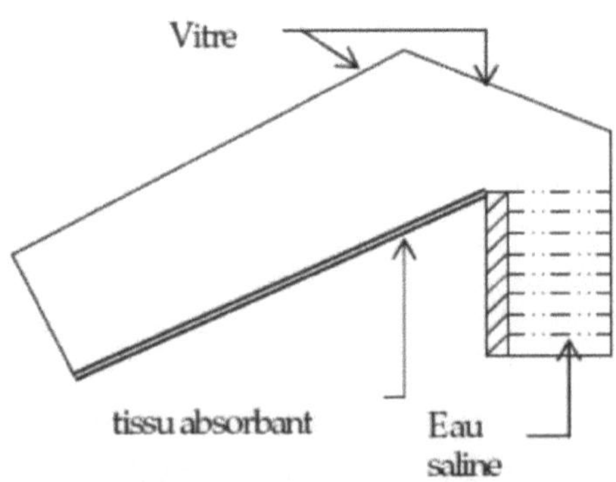

Figure 120 20- Self-powered distiller. [Source 5]

-Trickling distillers have a number of disadvantages, including :

- Difficult to adjust flow rate.
- A loss of sensible heat carried away by non-evaporated water.
- Significant use of water for distillation.

2-2-5-4-Vertical solar distiller :

The vertical solar distiller consists of:

- A solar collector to capture the sun's rays.
- An evaporation compartment.
- A condensation compartment.

Figure II 20 shows the operating principle of the vertical distiller. The distiller is fed with brackish water from a reservoir (1), the distiller is connected to the reservoir by means of a copper tube (2), the flow of water to the distiller inlet is regulated by means of a valve (3). The position of the feed tank allows the distiller to be fed at a constant flow rate.

A water dispenser (4), with perforations along its length, distributes the water in a film over the spongy fabric (5), the water runs off the fabric and the brine is drained off at the bottom (6).

The water that runs off behind the absorber (7) evaporates in the evaporation compartment, the mixture of water and air inside saturated with water vapour flows naturally through the two openings (8) made in an insulated partition towards the condensation compartment (9), and the vapour condenses on contact with the condensation plate at the rear of the distiller. Water collection occurs in the gutter (Figure **II-20**) fabricated below the condensation plate. **[27]**

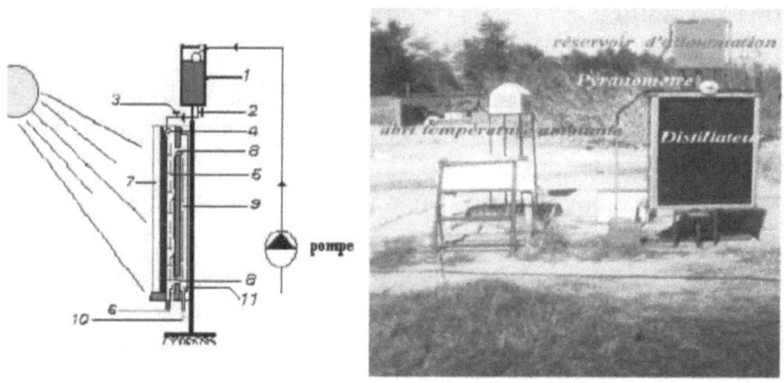

Figure II 21: Vertical solar distiller. [27]

2-2-5-5- Watercone distiller

This is a small inflatable solar distiller. Used to produce drinking water from seawater. [28]

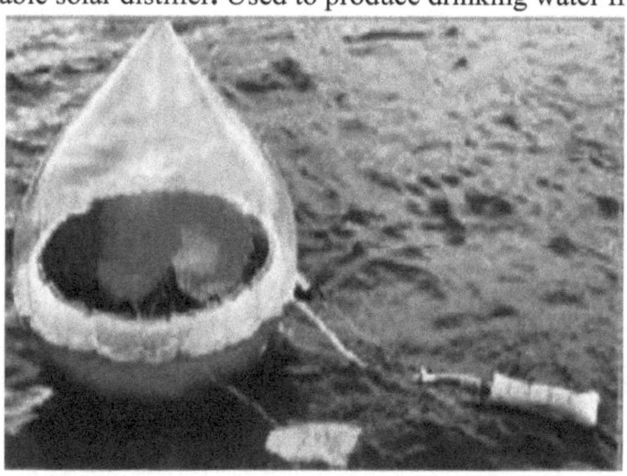

Figure II. 22: Inflatable solar distiller [28].

3-Solar distiller combined with other devices

Generally, the performance of solar distillers is low (2-5 l/m$^{(2)}$·day). [29], and also to ensure night-time production, complex systems with preheating of the water to be distilled or concentration of the energy are required. Without being exhaustive, we can mention :

3-1-Distiller with a solar thermal panel

The distiller can be coupled with a solar thermal collector. This principle is based on the idea of pre-heating the water to be distilled through thermal panels, which then emerges at a temperature of 70 to 80°C and enters the simple or multiple-effect distiller. In this type of desalination system, there are two modes of heat transfer fluid circulation:

- Natural circulation system (gravity circulation or thermosiphon).
- Forced circulation system.

3-1-1-Natural circulation system

The fluid circulates in closed circuits of temperature T_1 in one branch and T_2 in the other. In this very simple particular case, the fluid starts moving from the hottest parts towards the к"-most j. ices parts, the cl . ge Δp created by the difference in temperatures of the two branches TO A avir :

$$\Delta p = h(\rho(T_1) - \rho(T_2)) \qquad (2.1)$$

h: The difference in coastline.

ρ : Density of the fluid

- The equilibrium flow velocity of the fluid is obtained by writing that Δp is equal to the sum of the linear and local head losses in the circuit under consideration. This system must include a sensor and a reservoir raised above it.

In this case, the energy required for circulation is taken from the thermal energy supplied to the heat transfer fluid; the overall energy balance corresponds to that which would be obtained with a circulating pump. The advantage of this method of circulation is that there is no need for an electrical power supply; however, the circulation speed is low, which means that large pipe diameters are required, and the reverse effect, which is possible at night, requires precautions (non-return valves).

This system can use two types of storage.

3-1-1-1-Single-fluid system, storage without heat exchanger

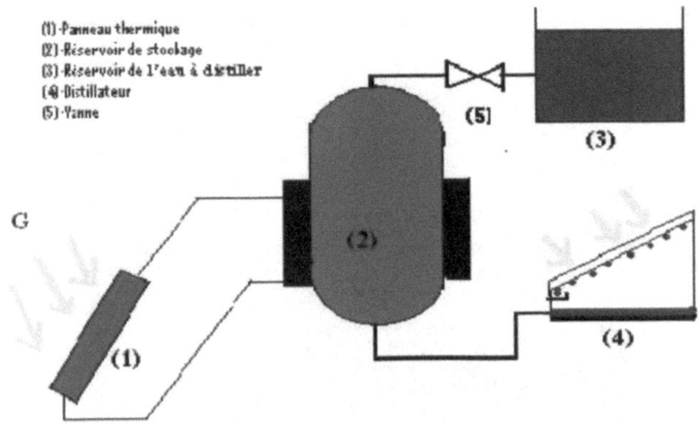

Figure II.23: Solar energy storage without heat exchanger. [Source 1]

This corresponds, in storage, to heat exchange by mixing, a system that is in any case more efficient than any tubular or plate heat exchanger. The major disadvantage of the system is that the fluid circulating in the collector is constantly renewed, which encourages the formation of insulating deposits on the inside of the absorber.

3-1-1-2-Two-fluid system, heat exchanger storage

The primary fluid (1) passes through an exchanger, usually tubular, placed at the base of the tank (R), and the secondary fluid (2) receives the heat from (1) via the exchanger. With this arrangement, a stratification of the water layers is established in (R) and the hot fluid ends up in the upper part where it can be drawn off; the water is topped up at the base of (R). The most widespread of the two systems is the forced circulation system with heat exchanger, as it offers the most flexible and safe operation. Month

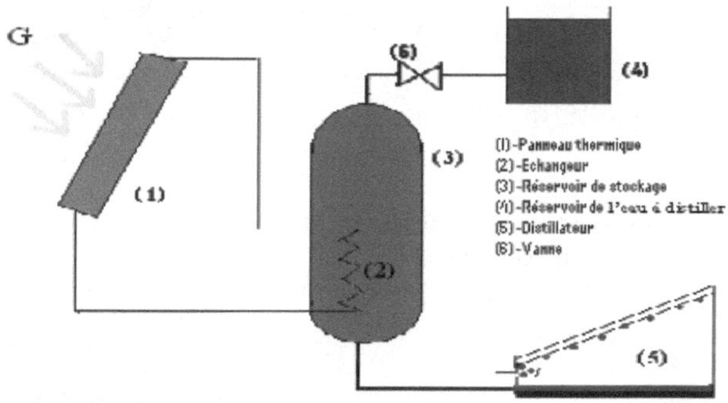

Figure II.24: Solar energy storage with heat exchanger. [Source 1]

3-1-2-Forced circulation system

Usually, much more flexible operation is obtained with a circuit comprising a P pump with adjustable flow rate. In this case, the reservoir can occupy any position, closer to the user in particular; in addition, high fluid circulation speed reduces the size of the pipes [29]. As before, this system can use two types of storage.

- single-fluid system, storage without heat exchanger.

Two-fluid system, heat exchanger storage.

3-2- Distiller with photovoltaic panel

3-2-1- Operating principle

The photovoltaic panel is used to convert solar energy into electrical energy, although the current is produced at variable voltages depending on the amount of sunlight.

For DC use, a voltage regulator is needed to charge a battery or accumulator (usually 12V). If the equipment is intended to operate on alternating current, an inverter will be required at the input to this equipment. The principle of a photovoltaic installation is illustrated in figure 23.

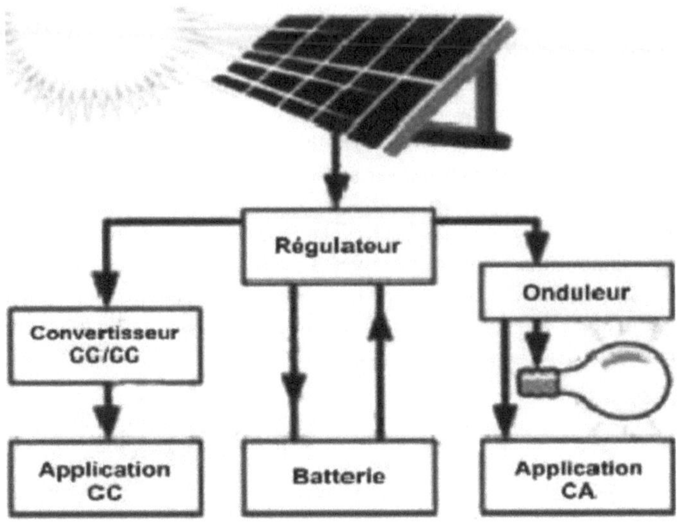

FigureII.25:Diagram of a desalination unit with a photovoltaic panel [Source1].

1-photovoltaic panel 2-regulator 3-exchanger 4-battery
5- tank
6-valve

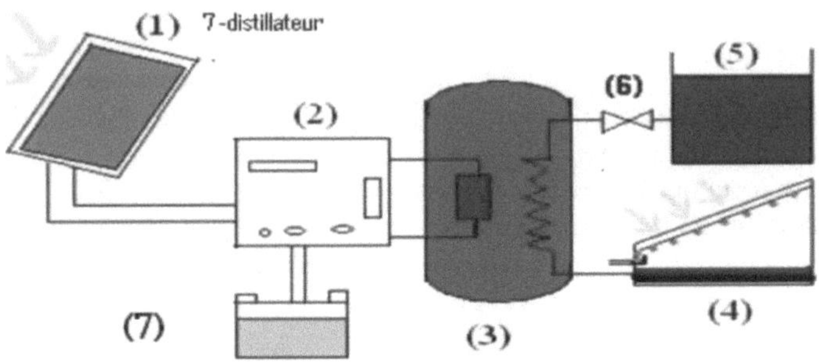

Figure II.26: Distiller with a storage tank preheated by
A photovoltaic panel. [Source 1]

The coupling of a distiller with photovoltaic panels uses resistors to heat the water entering the distiller to the desired temperature, see figure.

We can also heat a fluid (air, water) at the bottom of the distiller using electric resistors, as shown in the figure.

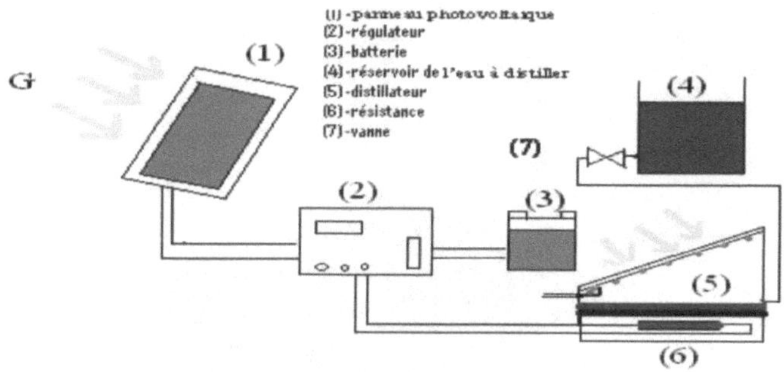

Figure II.27: Installation with photovoltaic panel and resistor At the bottom of the distiller. [Source 1]

3-3-Distiller with a wind turbine

A wind turbine can supply electricity to the grid and deliver significant power, of the order of 2 MW inland and 5 MW offshore.

Low-power wind turbines (typically 1 kW to 40 kW) are generally not connected to the electricity distribution networks, and the energy produced is intended to meet specific needs, for heating or electricity consumption.

3-3-1- The coupling principle

The distiller uses the heat produced by a resistor heated by the electricity generated by a low-power wind turbine (in the case of a wind turbine not connected to the electricity grid).

The principle is shown in Figure

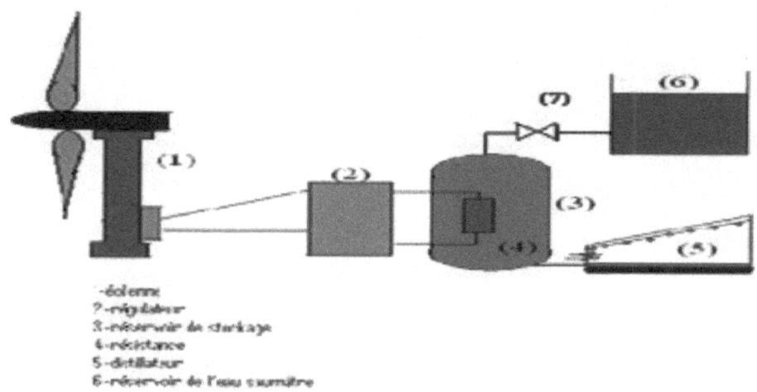

Figure II.28 : Coupling solar distiller and wind turbine. [Source 1]

3-4-Distiller with internal and external reflectors.

The use of internal or external reflectors (figure-25-), can be useful and inexpensive to increase the solar radiation incident on the cover of the pond, resulting in an increase in production.

Tamimi [30], reported that efficiency was increased by 20 to 30% when all the interior sides (sides and rear walls) were covered with reflectors.

-Experimental studies show that internal and external reflectors can remarkably increase distillate productivity all year round, except during the summer season.

Adding the internal and external reflectors to the standard single-slope distiller, the increase in daily quantities of distillate over the whole year is then brought to approximately 48% [31].

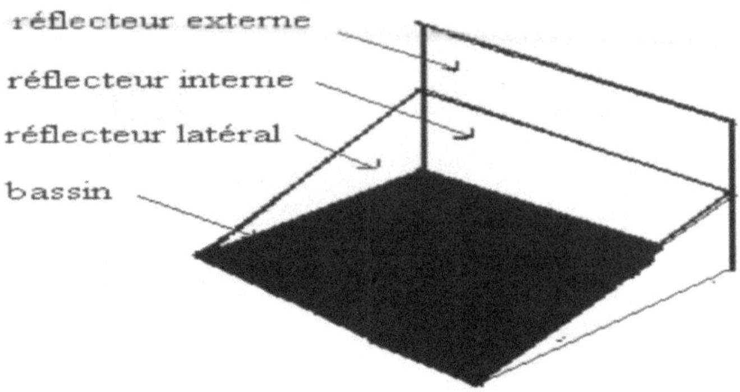

Figure 26 29: Solar distiller with reflectors
Internal and external. [Source 1]

-The bottom of the distillation unit is generally black to increase the absorption rate.

4-Operating parameters of a solar distiller

-It should be noted that **Satcunanathan** and **Hansen [32]** have defined a gross performance factor (GPF) and an hourly performance factor (HPF) which make it possible to characterise, in a simple manner, the performance of a machine.

plus absolute the distiller, their respective expressions are :

$$FPB = \frac{\text{quantté d'eau produite au bout de 24h}}{\text{quantité d'energie entrée au bout de 24h}} \qquad (2.2)$$

$$FPH = \frac{\text{quantité d' eau produite au bout d' une heure}}{\text{quantité d' energie entrée au bout d' une heure}}$$

-**At a** given moment of **the** day the performance factor F.P is given by the relation :

$$FP = \frac{m_d}{\alpha_t G A} \qquad (2.3)$$

P.F.: performance factor

5- Operating parameters and characteristics of a solar distiller

-Several variables are defined to characterise the production of water by such distillers. These include: yield, internal and overall efficiency, and performance. [24].

5-1-Yield :

This is the quantity of water produced per unit area of black plane per day. The major disadvantage of this criterion is that it does not me ion ck the solar energy that reaches the distiller.

The yield of a simple distiller is : $\eta = \frac{m_d h_{ev}}{G A}$ (2.4)

h_{ev}: Enthalpy of evaporation.

m_d : The distillate mass flow rate is calculated by :

$$m_d = \frac{q_{ev}}{L_v} \qquad (2.5)$$

G: Total solar energy incident per m^2 on a horizontal surface, over the course of a day.

5-2-Efficiency

The author [Zay 2002] defined седх types of efficiency:

5-2-1 Overall efficiency: η_g (°/₀) :

Representing the ratio of the amount of energy evaporated per (m2) to the total amount of energy incident per (m2) on a horizontal surface. It is defined by the ratio :

$$\eta_g = \frac{q_{ev}}{G \times A} = \frac{m_d \times L_v}{G \times A} \qquad (2.6)$$

With :

-q_{ev} : Quantity of heat used for evaporation per unit of time.

-A: Surface area of the collector (pane area).

-L_v : Latent heat of vaporisation.

5-2-2-Internal efficiency

While overall efficiency relates the quantity of water produced to the solar energy falling on a horizontal surface, it makes no mention of the quantity of energy actually entering the distiller, especially as a distiller is generally built for a given location and with a fixed roof slope. A change in slope and location will vary the amount of energy entering the distiller. We therefore define an efficiency which takes all these pai l metres into account and which

Is defined by :

$$\eta_i = \frac{Q_{ev}}{Q_{eau}} \qquad (2.7)$$

For an illumination intensity G, the expression Q_{eau} can be written as follows:

$$Q_{eau} = (\tau_v . \alpha_e + \tau_v . \tau_e . \alpha_b) . G . A \qquad (2.8)$$

Q_{eau} : Power absorbed by distilled water

Obviously, the heat absorbed by the bottom of the disti I teur is transferred entirely to the film of water by convection.

A coefficient known as the fictitious absorption coefficient can be defined for the body of water [33] by

posing :

$$Q_{eau} = \alpha_t . G . A \qquad (2.9)$$

The coefficient α_t is obtained using the optical and thermal characteristics of the distiller by the relation :

$$\alpha_t = \tau_v . \alpha_e + \tau_v . \tau_e . \alpha_b \qquad (2.10)$$

With

τ_v : the transmission coefficient of the glass

α_e: The water absorption coefficient i,

τ_e : The transmission coefficient of the glass

α_b : the absorption of the bottom of the distiller

So :

$$\eta_i = \frac{m_d L_v}{\alpha_t G A} \quad (2.11)$$

α_t : The overall absorption coefficient of the distiller.

The following table compares the different desalination technologies in terms of energy consumption and quantity of water produced.

6-Distiller time constant [34] **:**

The system time constant measures, in terms of time, the heat capacity of the system. It generally depends on the construction of the collector, the materials used, the heat of the heat transfer fluid, the heat extraction factor, the distillate flow rate and, of course, the operating temperature. A low response time has only a secondary effect on the collector's average long-term performance.

7- Parameters influencing the solar distillation system. [35]

The parameters influencing the solar desalination system can be classified into two types:

1 - External parameters

2 - Internal parameters

7-1- External parameters

These parameters are linked to the installation site of the desalination system

7-1-1-Geographical parameters

- Such as the longitude, latitude and height of the sun.

7-1-2-Meteorological parameters, the most frequently used of which are: **7-1-2-1-1Intensity of solar radiation :**

This is the essential factor in this study. It is radiant energy of short wavelength (0.17 to 4 μm), with maximum intensity at 0.47 μm in the visible spectrum. The overall intensity of solar radiation arriving at the edge of the atmosphere on a perpendicular surface is constant, equal to 1.35KW/m^2. However, the energy arriving near the Earth's surface is weakened by the absorption of part of the radiation by the atmosphere.

7-1-2-2-Wind speed, which plays a very important role, especially in heat exchange by convection between the external face and the environment. (There is a linear relationship between the heat exchange coefficient by conviction and wind speed).

7-1-2-3-Ambient temperature, an increase in which leads to a decrease in production. Other atmospheric parameters such as rain, dew and dust can also have an influence.

7-1-2-4 Cloud intermittence :

The amount of solar energy received by a surface depends on the parameters A and B, which depend on the nature of the sky. This means that the passage of clouds reduces the average effective absorptance.

7-2 Internal parameters

7-2-1-Position parameters

- The location of the distillation unit, where the distillation units must be placed so as to avoid obstacles, the "masking effect" that prevents solar radiation from reaching the collection surface:
- Orientation: this depends essentially on how the distiller operates during the day:
- Operation during the morning Eastbound ;
- Afternoon operation .... facing west;

-Operates throughout the day south-facing

- The inclination depends on the operation of the disti .hour during the year:

> -summer operation ....$\beta = \emptyset - 10$ (2.12)

> -winter operation. ... $\beta = \emptyset + 20$ (2.13)

> -annual operations... . $\beta = \emptyset + 10$ (2.14)

/> : The angle of inclination of the distiller glass

0 : The attitude of the place

7-3-Parameter of constrictions (geometric)

-The materials used in the construction of the distiller have a major influence on its operation, in particular :

7-3-1-Cover :

Coverage is provided mainly through :

-its nature

- its wettability by water
- its transparency to the sun's rays
- its opacity to long-wave infrared radiation
- its resistance to attack by wind and solid particles
- the number of panes if the roof is glazed.

- its inclination to the horizontal

7-3-2-The absorbent surface

-The role of the absorber is to transform the incident solar radiation into thermal flux. The ideal characteristics are :

-An absorption factor close to unity, good thermal conductivity and good chemical resistance to the fluid in which it is immersed.

-low thermal inertia. It must also be able to withstand high temperatures without damaging its surface. Aluminium and copper are usually coated with a thin layer of matt black paint, to increase its absorption capacity and reduce heat loss by reflection and diffusion.

7-3-3-The distance between the evaporating surface and the condensing surface

This parameter determines the intensity of lateral heat loss by convection, and the size of the buffer layer that slows down the exchange between the evaporating surface and the condensing surface. Research has shown that reducing this parameter increases the gross performance factor. [32]

7-3-4 Insulating the sides of the distiller :

-The purpose of insulating the sides is to eliminate losses (heat transfer with the outside environment).

7-3-5-Brine parameters :

- **M. Maalem** [36] observed that the greater the thickness of the water film inside the solar distiller, the less its temperature fluctuates during production. Thermal inertia increases with thickness, and night-time operating time also increases with thickness, although average production over 24 hours decreases with increasing thickness. Experiments have shown that this production follows the law :

$$y = \frac{a}{x+c} + b \qquad (2.15)$$

With:

Y: Production

X: The thickness of the brine

a, b, c: Parameters depending on the shape of the device and local conditions.

-Brine temperature: experiments carried out in the Algerian Sahara on greenhouse solar distillers have shown that the instantaneous flow rate as a function of temperature is calculated by the following correlation:

$$y = \frac{51.607}{117-T} - 0.613 \qquad (2.16)$$

With :

Y: The instantaneous flow rate, expressed in litres per hour.

T: The temperature, in degrees centigrade, of the brine in the upper layer.

-When the temperature of the brine increases, a white deposit on the free surface of the water due to the formation of carbonates insoluble in the brine has been observed; the salt deposit acts on the absorbing power of the black surface and considerably reduces production.

- Salt concentration: The distiller's output decreases as the concentration of salt in the water to be distilled increases.

7-3-6- Optical parameters :

The parameters are the emissivity, absorptivity, reflectivity and transmittance of the absorbing surface and the blanket.

7-3-7-Thermophysical parameters :

The thermo-physical parameters are :

- The thermal conductivity, specific heat and thermal diffusivity of the roof, and the absorbing surface.
- the thermal conductivity, specific heat, kinematic viscosity, latent heat of vaporisation, coefficient of thermal expansion and thermal diffusivity of the brine.
- The thermal conductivity, specific heat, dynamic viscosity, kinematic viscosity and coefficient of thermal expansion of the air-steam mixture.

8- Objective design for an efficient solar distiller :

For high efficiency, the solar distiller should maintain :

- High feed water temperature.
- A large temperature difference between the supply water and the condensation surface.
- Low steam leakage.
- A large temperature difference can be made if :
- The surface of the condensation absorbs little or no incoming radiation.
- Condensation absorbs heat, which must be quickly removed from the unit.

The condensation surface.

In addition, the following elements must be monitored:

- Location

Solar distillers operate using direct solar energy without the need for additional power.

- Consumption and additional space

Distilled water from the distiller can be warm to hot. It is best to refrigerate the water in a glass jug to cool it and store it.

- **Taste**
- Distilled water is tasteless due to the absence of minerals, according to Personal Preferences. Cooling the water may improve the taste.
- **Easy to clean**

Choose a model with an easy-to-clean absorber.

- **The composition of distillers**

We recommend that distillers be made of galvanised iron, aluminium or plastic.

- The following table compares the different desalination techniques

The desalination technique	Desalination system characteristics	Energy consumed (kWh/m3)
Osmosis reverse	-low production: 500 to 5000 m3 /day - applied to brackish water (salinity less than 10 g/l) - separation of organic molecules $10^{(-7)}$ mm in size Water quality with a salinity of around 0.5 g/l	4-5
Electrodialysis	- the amount of energy consumed depends on the salt concentration. - a small quantity of water produced - brackish water with a salinity of less than 10g/l	5
Vacuum distillation	very pure water	2-3
Multi-Stage (Flash distillation MSF	- large capacity: over 5000 m3/day - seawater with a salinity greater than 30 g/l	15
Multi-Effect (MED distillation).	- large capacity - sea water	15
Desalination vacuum	-large capacity - seawater	5
Desalination by freezing	- large capacity - sea water	5
Solar distillation	2250 kj (if the change of state takes place at 100°C) - very economical method - relatively low yield: 4 to 5 $l/m^{(2)}$ ·day	Free energy is clean

Table II.1: The different desalination methods. [11]

9- State of the art in solar distillation :

The complexity of modelling heat and mass transfer within different types of solar stills has led to the proposal of several mathematical models to govern the operation of these devices. We present in the following, some results of scientific work identified in the literature.

The first in the mathematical modelling of a solar distiller is achieved by **Dunkle** who to present a complete mathematical formulation and a fundamental theoretical model for the prediction of trafori ce chalo processes. <ι c.c mass in solar distillatemi. This analysis was based on the deso юн-! du ermic by free convection based on the dimensionless correlation. Λ 3.075. Λ' . I on the upward heat flux in the horizontal spaces. [37]

Rheinlander proposed a mathematical model for the numerical solution of heat and mass transfer equations in solar systems. The results obtained were successfully compared with the early work of Cooper, Kumar and Tiwari [38].

A simplified analysis was presented by **Malik et al.** This analysis, based on more simplified assumptions than those of Dunkle, leads to the basic correlations of heat and mass transfer, which were deduced from the Lewis relation [39].

Kumar and Tiwari, Tiwari et al, Aggrawal and Tiwari have obtained new and improved values of the numerical constants in the dimensionless correlation of the Nusselt number [40] [41] [42].
Tiwari et al obtained better correlation parameters for heat transfer models by measuring the temperature of the internal surface of the glazing [43].
Tripathy and Tiwari studied the influence of the depth of the brine layer on the heat and mass transfers that occur in either passive or active distillers and also the effect of the inclination of the condensing surface on the condensation processes. [44].

Chapter 3

Modelling and Numerical Resolution

-1- Introduction

In this part of the work, the theoretical study of the double-slope solar distiller will be presented. The operating principle of the double-slope solar distiller will be discussed, and an attempt will be made to establish the heat balance for each part of the distiller and the resulting system of equations using a numerical approach based on the fourth-order **range kutta** method.

2-The operating principle of a double-slope solar distiller

A solar distiller consists of a watertight container topped by a glass pane. The lower part is covered with a body of water (brackish water); the water evaporates under the action of the incident solar flux, condenses on the glass and is then collected in a receiver (the same principle as a simple solar distiller except that this type of distiller contains two panes with the same angle of inclination). A water make-up tank must compensate for the flow of distillate. So we have a hot source (the mass of water to be distilled) and a cold source (the glass). So we have an exchange of heat between these two sources.

3-Heat exchange in a double-slope solar distiller

Before establishing the overall energy balance of a solar distiller, it is first necessary to determine the main heat transfers that occur within the distiller.

3-1-Window level

On the inside, the pane receives flows by convection **qci**, radiation **qri** and evaporation **qe.**

- On the outside, the glass gives off the sum of two flows **qca** , by convection and **qra** , to the ambient air. By radiation

3-2-Brine level

The brine transfers the **qri** flow by radiation, the **qci** flow by natural convection and **the qe** flow by evaporation (during phase change) to the condensation surface (inside of the distiller window).

3-3- In the absorber :

The sheet of water receives a flow of heat **qeb** from the absorber (tank) by conduction and convection on the one hand, and gives off a flow **perts** to the outside by conduction on the other.

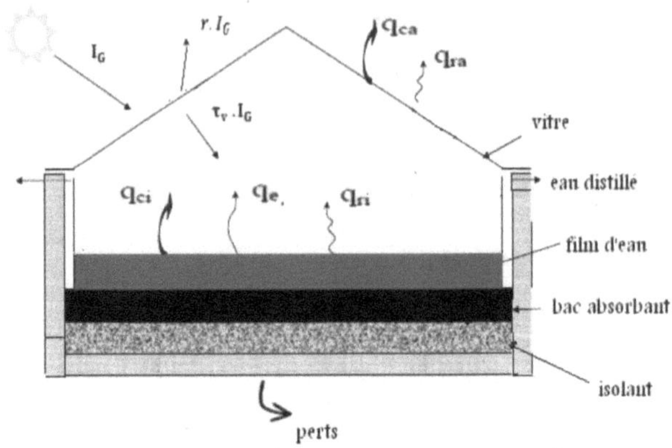

Figure -III-1: The various heat transfers in a solar distiller

4-Mathematical modelling of heat exchange in a solar distiller

4-1-Simplifying assumptions

Simplifying assumptions have been adapted, namely :

- the celestial vault is considered to be a black body.
- the side walls are assumed to be well insulated.
- dissolved salts have no influence on the amount of heat exchanged by the brine.
- The salt water in the tank is static
- The temperatures of the glass, the water film and the absorber are uniform
- Heat losses due to the extraction of the distillate are neglected.
- The thermal inertia of the roof is low.

Condensation occurs only on the roof (there is no condensation on the side walls).

4-2-Thermal balance of the double slope solar distiller for a steady state The theory of a double slope solar distiller is presented by **Mousa abu-Arabi** et **al** [44], [45] and **Yousef. H** [46], and **B. Bouchekima** [47].

- We consider any section of the system at time "**t**", let **i** be one of the media represented in this section **mi,** its mass in [Kg], **CP** its specific heat in [j/Kg °C], T its temperature in [°C]. The balance at node **i** gives

$$m_i C_{pi} \frac{dT}{dt} = \sum Q_{ij} + P_i \qquad (3.1)$$

Pi: Term for spring or well.

$$m_i C_{pi} \frac{dT}{dt} = 0 \qquad (3.2)$$

In the following cases:

-A distiller with negligible thermal inertia (low mass of its components or low specific heat).

-A stationary regime

4-2-1-Thermal balance of the pane

$$m_v C_{Pv} \frac{dT_v}{dt} = P_V + q^c_{e-v} + q^r_{e-v} + q^{ev}_{e-v} - q^c_{v-a} - q^r_{v-a} \qquad (3.3)$$

q^{c}_{e-v} : Heat flux exchanged by convection between the inner pane and the water film **[W].**

$$q^{c}_{e-v} = h_{ci} * (T_e - T_v) \quad (3.4)$$

T_v : The temperature of the glass.

T_e : Water temperature

q^{r}_{e-v} : Heat flux exchanged by radiation between the pane and the body of water **[W]**

$$q^{r}_{e-v} = h_{ri} * (T_e - T_v) \quad (3.5)$$

q^{ev}_{e-v} : Heat flux exchanged by evaporation between the glass and the water mass **[W].**

$$q^{ev}_{e-\epsilon} = h_{ev} * (T_e - T_v) \quad (3.6)$$

q^{c}_{v-a} : Heat flux exchanged by radiation between the ambient environment and the external surface of the pane **[W].**

$$q^{c}_{v-a} = h_{ca} * (T_v - T_a) \quad (3.7)$$

q^{r}_{v-a} : Heat exchanged by convection between the ambient environment and the external surface
of the glass ***[W]***

$$q^{r}_{v-a} = h_{ra} * (T_v - T_a) \quad (3.8)$$

P_V : Solar power absorbed by the glass

$$P_v = I_D \times a_{vD} + I_d \times a_{vd} \quad (3.9)$$

Or:

a_{vD} and a_{vd} : are the absorption coefficients respectively for direct radiation and diffuse.

$$a_{vD} = 1 - R_D - T_D \quad (3.10)$$

Or:

T_D : The overall transmission coefficient of direct radiation is calculated by :

$$T_D = T_r \times T_a \quad (3.11)$$

T_r : The transmission coefficient due to reflection

$$T_r = (1 - R_D) / (1 + R_D) \quad (3.12)$$

T_a : The transmission coefficient due to absorption

$$T_a = exp\,[-k_v \times e_v / cos\,(\theta)] \quad (3.13)$$

Or:

k_v : Glass extinction coefficient $[m^{-1}]$

e_v : Glass thickness $[m]$

R_D Characterises the reflectivity of the glass and is given by :

$$R_D = (R_N + R_p)/2 \quad (3.14)$$

With

$$R_N = \frac{\sin^2(\theta - i)}{\sin^2(\theta + i)} \quad et \quad R_P = \frac{tg^2(\theta - i)}{tg^2(\theta + i)} \quad (3.15)$$

Or:

θ: Angle of refraction of the ray at the pane.

i and θ_r are linked by the Fennell-Descartes relationship:

$$n_1 \times \sin(i) = n_2 \times \sin(\theta) \quad (3.16)$$

With

n_1 : Absolute refractive index of the medium through which the incident ray passes

n_2 : Absolute refractive index of the medium through which the refracted ray passes

The absorption coefficient for diffuse radiation is calculated in a similar way to that for direct radiation, with an angle of incidence of $i = 60°$.

$$a_{vd} = 1 - R_d - T_d \quad (3.17)$$

4-2-2-Thermal balance of the water body

$$m_e C_{pe} \frac{dT_e}{dt} = P_e - q^c_{e-v} - q^r_{e-v} - q^{ev}_{e-v} + q^c_{b-e} \quad (3.118)$$

q^c_{b-e} : Heat flux exchanged by convection between the absorber and the water film **[W]**.

$$q^c_{b-e} = h_{be} * (T_b - T_e) \quad (3.19)$$

P_e : Solar power absorbed by the water film

$$P_e = I_D \times a_{eD} + I_d \times a_{ed} \quad (3.120)$$

a_{eD} and a_{ed} : are respectively the brine absorption coefficients for the direct and diffuse radiation.

$$a_{eD} = T_D \times (a_e + T_{Re} \times a_b) \quad (3.21)$$

With :

a_e : Brine absorption coefficient (constant)

T_{Re} : Brine transmission coefficient

$$T_{Re} = a_e - R_e \quad (3.22)$$

R_e : The reflection coefficient of brine is given by :

$$R_e = R_N + R_P \quad (3.23)$$

Or:

a_b: Absorption coefficient of the absorbent tray.

The absorption coefficient for diffuse radiation a_{ed} varies with respect to a_{eD}; andte variation is due to T_d.

4-2-3-Thermal balance of the sorbent tank

$$m_b C_{pb} \frac{dT_b}{dt} = P_b - q^c_{b-e} - q_{pert} \quad (3.24)$$

$$q_{pert} = U_b (T_b - T_a) \quad (3.25)$$

P_b : Solar power absorbed by absorber **[W]**

If α_b is the absorption coefficient of the absorber tray, we can deduce the power : P_b:

$$P_b = \frac{I_D \times T_p + I_d \times T_d}{1-(1-\alpha_b) \times R_d} \quad (3.26)$$

$$q_{perts} = U_b * (T_b - T_a) * A_b \quad (3.27)$$

According to **H.Hilal** et **al** [48], the overall loss coefficient is given by the formula :

$$U_b = (\frac{1}{h_b} + \frac{\delta}{e})^{-1} \quad (3.28)$$

With :

δ, e : Are the conductivity and thickness of the insulating layer respectively

Condensate flow rate

$$\frac{dM_c}{dt} = h_{evap} \cdot \frac{(T_e - T_v)}{h_{fg}} \quad (3.29)$$

M_c : Condensate mass.

h^{ev}: The latent heat of vaporisation.

5-Expression of the various heat exchange coefficients :

5-1-the sun's rays

Solar radiation is characterised by various quantities, essentially the total energy sent by the sun to the edge of the Earth's atmosphere, known as the "solar constant", and the various (random) alterations and attenuations that solar radiation undergoes as it passes through the Earth's atmosphere.

5-1-1- The solar constant

The solar constant is defined as the total energy sent by the sun to the boundary of the earth's atmosphere over a surface area of 1m^{2}placed perpendicular to the direction of the radiation; it is independent of weather conditions, but depends on the earth-sea distance e i

$$C^{te}\ solaire = I_0 = \sigma \times T_s^4 \times (\frac{R_s}{D_{st}})^2 \quad (3.30)$$

With :

R_s: The sun's ray (m)

D_{st}: distance earth-sun

$\sigma 1$: Stéphane-Boltzmann constant

T_s: the temperature of the sun

As the distance between the sun and the earth varies over the year, the average value of the solar constant (I0) is of the order of 1367 (w/m2). This figure is generally corrected by the following relationship:

$$I_C = I_0 \times (1 + 0.033 \times cos\ (\frac{360}{365} \times d)) \quad (3.31)$$

Or by declination:

$$I_c(\delta) = I_0 \times (1 - \frac{\sin(\sigma)}{11.7}) \quad (3.32)$$

Where:

d: declination (°)

n: number of the day in the month

m: number of the month in the year

d: Number of the day in the year (table)

Month	jan	Feb	March	April	May	June	July	August	Sep	Oct	Nou	dec
Number From the 1st day Of each month in the year	1	32	60	91	121	152	182	213	244	274	305	335

Table-III -1- The day of the year. [BOU 2009]

5-1-2-Solar radiation at ground level

Solar radiation undergoes a number of random alterations and attenuations as it passes through the atmosphere; reflection off atmospheric layers, molecular absorption, molecular scattering and by aerosols (dust, droplets....)

At ground level, because of scattering, part of the radiation is diffuse (i.e. isotropic). The other, direct, part is anisotropic. Overall radiation is therefore the sum of direct and diffuse radiation.

5-1-3- Calculation of solar radiation

The solar radiation received by the glass pane can be measured either directly using a pyranometer fixed to a support inclined at an angle equal to the latitude of the site of the experiment or by empirical formulae as follows:

5-1-4- Direct radiation, which comes from the solar disc and is attenuated as it passes through the atmosphere. The expression for direct radiation for a plane facing south is given by [26], [49], [50], [51], [52].

$$I_D = I_C \times a \times exp\left(\frac{-b}{sin(h)} \times \frac{p}{1000}\right) \times cos(i) \quad (3.33)$$

$$I_C = I_0 \times (1 + 0.034 \times (cos\,(30 \times (m-1) + n) \quad (3.34)$$

a, b: Coefficients reflecting atmospheric disturbances.

Table 2 shows some values for a and b.

$$sin(h) = cos(\sigma) \times cos(\omega) \times cos(\emptyset) + sin\,(\sigma) \times sin\,(\emptyset) \quad (3.35)$$

h: Eü.r.t the height of the sun

$w, \emptyset, \delta$: These are the hour angle, declination and latitude (°) respectively.

The atmospheric pressure "p" as a function of altitude is shown in the table :

$$cos(i) = cos(\omega) \times cos(\sigma) \times cos\,(\emptyset - \beta) + sin\,(\sigma) \times sin\,(\emptyset - \beta) \quad (3.56)$$

Where β is the inclination of the plane under consideration

	Clear skies	Condition normal	Industrial estates
A	0.87	0.88	0.91

B	0.17	0.26	0.43

Table-III-2-Values of atmospheric disturbance coefficients.

Attitude	0	500	1000	1500	2000	5000
P (mbar)	1000	950	900	850	800	500

Table-III-3-Variation of atmospheric pressure with attitude

5-1-5- Diffuse radiation comes from all the elements irradiated by the sun. The diffuse flux received by the sensor surface is characterised by two flux components: The diffuse flux emitted by the sky (φ_1), , given by the relation/

$$\varphi_1 = \varphi_0 \times \frac{1+\cos(\beta)}{2} \qquad (3.37)$$

The flux emitted by the soil φ2 by the relation :

$$\varphi_1 = \rho \times \frac{1-\cos(\beta)}{2} \times (I_c \times \tau_{dir} \times \sin(h) + \varphi_0) \qquad (3.38)$$

With:

$$\varphi_0 = I_c \times \sin(h) \times (0.271 - 0.2939 \times \tau_{dir}) \qquad (3.39)$$

τ_{dir} : The total transmissivity of the atmospheric layer, expressed as :

$$\tau = a \times exp\left(\frac{-b}{\sin(h)} \times \frac{P}{1000}\right) \qquad (3.40)$$

ρ The albedo of the sole is given in the following table:

soil type	Albedo
Fresh grass	0.20
Dry grass	0.30
Bare cultivated land	0.16
Gravel	0.22
Fresh snow	0.85
Stretch of water	0.05

able-III-4-albedo of different soils.

The total diffuse flux can therefore be written as follows:

$$I_d = \varphi_1 + \varphi_2 \qquad (3.41)$$

5-1-6- Global radiation: Global radiation on any surface near the ground is the sum of these two components:

$$I_G = I_D + I_d \qquad (3.42)$$

5-2-Expressions of different heat exchange coefficients

5-2-1-Water-to-glass convection

$$q^c_{e-v} = h^c_{e-v} * (T_e - T_v) \qquad (3.43)$$

h^c_{e-v} : The heat transfer coefficient between the water film and the glass is given by the relationship [44] :

$$h^c_{e-v} = 0.884[T - T + \frac{(P_e - P_v)\cdot(T_e + 273.15)}{268.9.10^3 - P_e}]^{1/3} \qquad (3.44)$$

P_e : The partial pressure of water at Te

P_v : The partial pressure of the glass at Tv

3-5-2-2-By evaporation :

$$q^{ev}_{e-v} = h^{ev} * (T_e - T_v) \qquad (3.45)$$

h^{ev} : The evaporative heat transfer coefficient between the film of water and the glass is given by:

$$h^{ev} = 16.273.10^{-3} * h^c_{e-v} * \frac{(P_e - P_v)}{(T_e - T_v)} \qquad (3.46)$$

The pressure of water vapour and the latent heat of evaporation are estimated by the following relationships [Fem 1990]:

$$P_i = exp\,(25.317 - \frac{5144}{T_i + 273.15}) \qquad (3.47)$$

Another linear relationship between pressure and temperature (T in C°) is given as follows [Kam 1987] :

$$P = 0.14862\cdot T - 0.36526\cdot 10^{-2}\cdot T^2 + 0.11242\cdot 10^{-3}\cdot T^3 \qquad (3.48)$$

$$h_{fg} = 3.1615 * 10^6 - 2.40714 * 10^3 T \qquad (3.49)$$

5-2-3-Water-to-glass radiation

$$q^r_{e-v} = h^r_{e-v} * (T_e - T_v) \qquad (3.50)$$

The formula for the coefficient of heat transfer by radiation h^r_{e-v} ' between the film of water and the glazing is given by :

$$h^r_{e-v} = \frac{\varepsilon_{eff} * \sigma 1 * [(T_e + 273.15)^4 - (T_v + 273.15)^4]}{T_e - T_v} \qquad (3.51)$$

ε_{eff} : Effective emissivity.

$$\varepsilon_{eff} = (\frac{1}{\varepsilon_e} + \frac{1}{\varepsilon_v} - 1)^{-1} \quad (3.52)$$

ε_e : Water emissivity

ε_v : Emissivity of the glass

$\sigma 1$: The Stefan-Boltzmann constant

5-2-4-Radiation from window to room :

$$q^r_{v-a} = h^r_{v-e} * (T_v - T_a) \quad (3.53)$$

$$h^r_{v-r} = \frac{\varepsilon_v * \sigma 1 * [(T_v+273.15)^4 - (T_{ciel}+273.15)^4]}{T_e - T_v} \quad (3.54)$$

T_{ciel} : The temperature of the sky

-The temperature of the sky estimated by the correlations given below [46] and [48].

$$T_{ciel} = T_{air} - 6 \quad (3.55)$$

$$T_{ciel} = T_{air} - 12 \quad (3.56)$$

-In general, we use the expression :

$$T_c = 0.0552 * (T_a)^{1.5} \text{ , } T_a \text{ en K}^{\circ} \quad (3.57)$$

5-2-5-Convection with ambient glass :

$$q^c_{v-a} = h^c_{v-a} * (T_v - T_a) \quad (3.58)$$

The convection exchange coefficient between the external face and the ambient environment is given by the following relationship

$$h^c_{v-a} = 5.7 + 3.8 * v \quad (3.59)$$

A second correlation is given by [45] :

$$h^c_{v-a} = 2.8 + 3.0 * v \quad (3.60)$$

A third correlation is given by [47] :

$$h^c_{v-a} = 2.8 + 3.0 * V, \; v \le 5\,(m/s) \quad (3.61)$$

(3.52)

$$h^c_{v-a} = 6.15 * v^{0.8}, v > 5\,(m/s) \quad (3.62)$$

Where is the wind speed?

5-2-6-Water basin convection

$$q^c_{b-e} = h^c_{b-e} * (T_b - T_e) \quad (3.63)$$

The convective heat exchange coefficient between the water film and the absorber tray is given by :

$$h^c_{b-e} = (Nu * \lambda_e)/L \quad (3.64)$$

λ_e : Thermal conductivity of water

-In this type of problem, experimental correlations lead to relationships of the form [Bac 1999], [Tchinda.2004] :

$$Nu = C * (Gr * Pr)^n \quad (3.65)$$

Where **c** and **n** are constants that depend on the geometry of the system and the nature of the laminar or turbulent flow

Gr: nombre de GRASHOF , $Gr = \frac{B_e \cdot L^3 \cdot g \cdot (T_b - T_e)}{\nu^2}$ (3.66)

B_e: Coefficient of thermal expansion of **water**

L: Length of absorbent tray

Pr: Prandtl number

Pr : Nombre de Prandtl, $Pr = (\mu_e * C_{pe})/\lambda_e$ (3.67)

- $Gr < 10^5 , Nu = 1$ (3.68)

-Heat exchange between the brine and the absorbent tank is solely by conduction.

- $10^5 < Gr < 2 * 10^7 , Nu = (\lambda_e/L) * 0.54 * (Gr * Pr)^{0.25}$ (3.69)

- $Gr > 2 * 10^7 , Nu = 0.14 * (Gr * Pr)^{0.33}$ (3.70)

h_b : The convection heat transfer coefficient between the absorber tray and the outside.
hb is calculated using the equations above.

6-Resolving the system of equations

The equations governing the operation of the double-slope solar distiller in this case consist of four 1st order differential equations. We need to determine the four unknowns: Tv, Te, Tb, Mc

To solve this system of equations we have used a numerical approach based on on the 4th order RANGE-KUTTA method, because it is the most widely used in practice. To do this, we begin by ordering the final equations of the various balances with respect to temperature as follows:

On the glass :

$$m_v C_{pv} \frac{dT_v}{dt} = P_v + (h^c_{e-v} + h^{ev}_{e-v} + h^r_{e-v}) * (T_e - T_v) - (h^r_{v-a} + h^c_{v-a}) * (T_v - T_a) \quad (3.71)$$

Water film level

$$m_e C_{pe} \frac{dT_e}{dt} = P_e + h^c_{b-e} * (T_b - T_e) - (h^c_{e-v} + h^{ev}_{e-v} + h^r_{e-v}) * (T_e - T_v) \quad (3.72)$$

Absorbent tray

$$m_b C_{pb} \frac{dT_b}{dt} = P_b - h^c_{b-e} * (T_b - T_e) - U_b * (T_b - T_a) \quad (3.73)$$

Condensate flow rate

$$\frac{dM_c}{dt} = h_{evap} * \frac{(T_e - T_v)}{h_{fg}} \quad (3.74)$$

7-Computer processing

The computer program developed is written in FORTRAN (**Compaq Visuel Fortran Version 6.6**) and essentially contains a main program and four subroutines. This software calculates the flow received by the glass every hour, the different heat flows exchanged (by convection, radiation, evaporation, conduction), the variation in the temperatures of the different elements of the distiller, the condensate rate, the overall efficiency and the internal efficiency of the distiller and finally the performance factor.

The main program calculates the solar irradiance every hour and links the sobroutines.

Sobroutines are used to calculate: glass temperature, water temperature, condensate level, etc. Before starting the programme, you are given meteorological data, such as the initial temperatures (glass, water, absorbent tray), solar irradiance, wind speed, the latitude of the location, and the initial and final times of the calculation, and optical properties such as the absorptivity of the glass and the water in the absorbent tray.

8-Organigramme

8- Organigramme

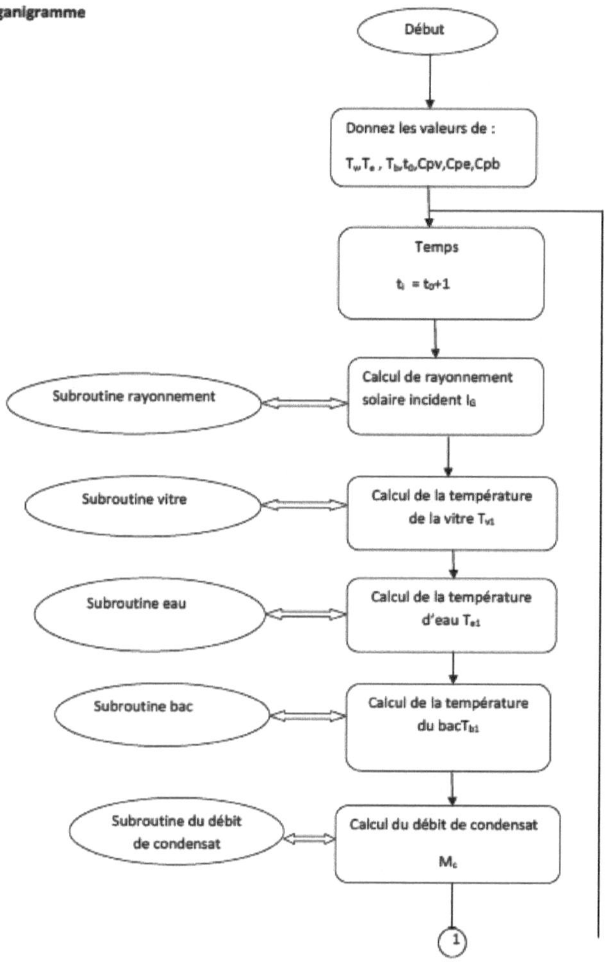

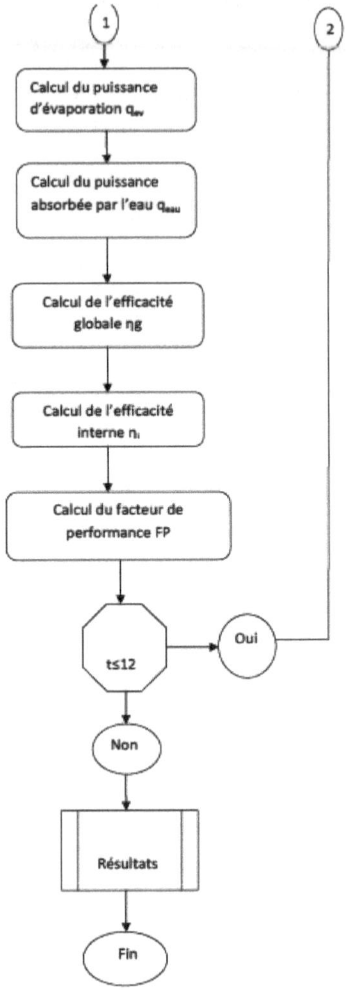
1
2
Calcul du puissance d'évaporation q_{ev}
Calcul du puissance absorbée par l'eau q_{eau}
Calcul de l'efficacité globale ηg
Calcul de l'efficacité interne η_i
Calcul du facteur de performance FP
t≤12
Oui
Non
Résultats
Fin

Organigramme de calcul

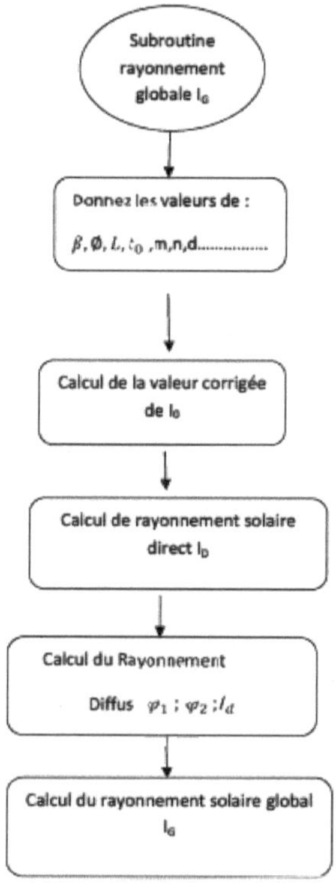

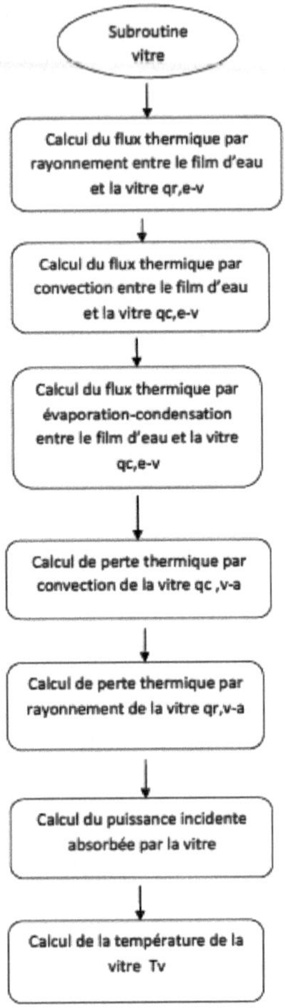
Subroutine vitre
Calcul du flux thermique par rayonnement entre le film d'eau et la vitre qr,e-v
Calcul du flux thermique par convection entre le film d'eau et la vitre qc,e-v
Calcul du flux thermique par évaporation-condensation entre le film d'eau et la vitre qc,e-v
Calcul de perte thermique par convection de la vitre qc ,v-a
Calcul de perte thermique par rayonnement de la vitre qr,v-a
Calcul du puissance incidente absorbée par la vitre
Calcul de la température de la vitre Tv

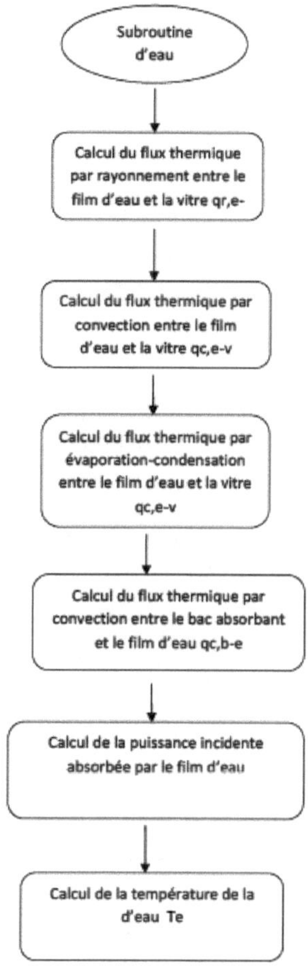
Subroutine d'eau
Calcul du flux thermique par rayonnement entre le film d'eau et la vitre qr,e-
Calcul du flux thermique par convection entre le film d'eau et la vitre qc,e-v
Calcul du flux thermique par évaporation-condensation entre le film d'eau et la vitre qc,e-v
Calcul du flux thermique par convection entre le bac absorbant et le film d'eau qc,b-e
Calcul de la puissance incidente absorbée par le film d'eau
Calcul de la température de la d'eau Te

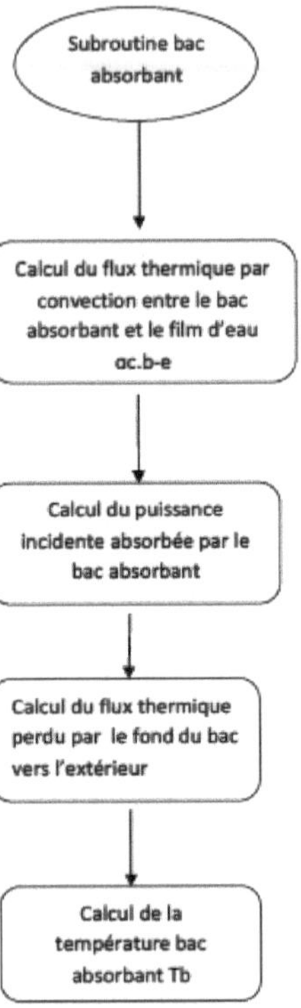
Subroutine bac absorbant
Calcul du flux thermique par convection entre le bac absorbant et le film d'eau αc.b-e
Calcul du puissance incidente absorbée par le bac absorbant
Calcul du flux thermique perdu par le fond du bac vers l'extérieur
Calcul de la température bac absorbant Tb

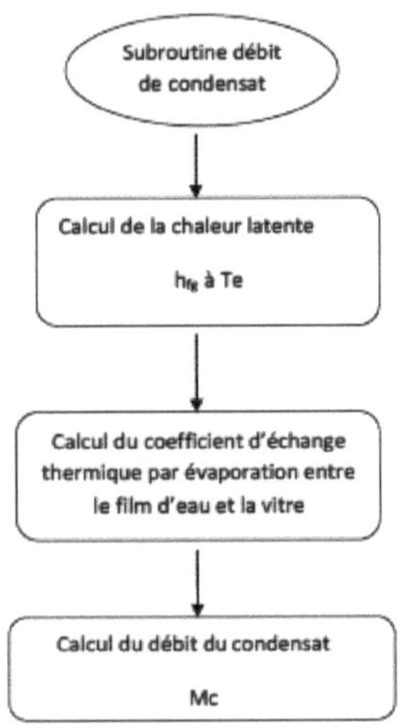
Subroutine débit de condensat
Calcul de la chaleur latente
h_{fg} à Te
Calcul du coefficient d'échange thermique par évaporation entre le film d'eau et la vitre
Calcul du débit du condensat
Mc

Chapter 4

Results and interpretation

1-Introduction

The theoretical optimisation of the performance of the double slope solar distiller in a desert environment is the main objective of this work. Theoretical results are validated by compression with experimental results.

The experimental results of this work is carried out during the day 19 April, in the city of Ouargla which is located at a latitude of 31.57 ^{0}North, and a longitude of 5.24 East, and an altitude of 135 m at sea level, and a time difference of one hour.the tests conducted on a solar distiller with double slope having the following properties (the results found by the student douadi rachida for the realization of his magister work under the title

دراسة مقطر شمسي ذي مكثف مظلل جزئيامقطر شمسي ذي مكثف مظلل جزئيا،بفعل الاحتباس الحراري الشمسي في المناطقالقاحلة) .

Absorber: Aluminium material,

Thickness: 0.03 m

Length: 90 cm

Width: 70 cm

Surface area: 0.63 m2

A glass cover :

Length: 0.94 m

Width: 0.40 m

Thickness: 0.004 m

Tilt to the horizontal: 10 degrees on both sides with an azimuth of 0 degrees to the south

Water depth: 1.5 cm

insulation: polystyrene

Thickness: 0.06 m

Base: galvanised material

Thickness: 0.5 mm

The theoretical results found by the calculation code are presented in the form of graphs plotted by L'ORIGINE with the experimental results. These graphs essentially concern the variation of the global illuminance incident on the glass at different angles of inclination, the different temperatures in the solar distiller during the day of the experiment, the daily production, the internal and global efficiency and, finally, the performance factor.

2-Variation in solar radiation for different atmospheric haze conditions

The following figure shows the variation in total solar radiation received by the glass surface of the double-slope solar distiller with an inclination of $10^0$to the horizontal, according to different atmospheric haze coefficients (normal conditions, clear sky, and an industrial zone). The variation of the solar flux incident on the glass follows the shape of a bell. It is more intense at dawn, while it is less intense at dusk. The maximum values of solar radiation are found for a pure sky where all the radiation passes through the atmospheric layer towards the earth without undergoing any deviation or deformation.

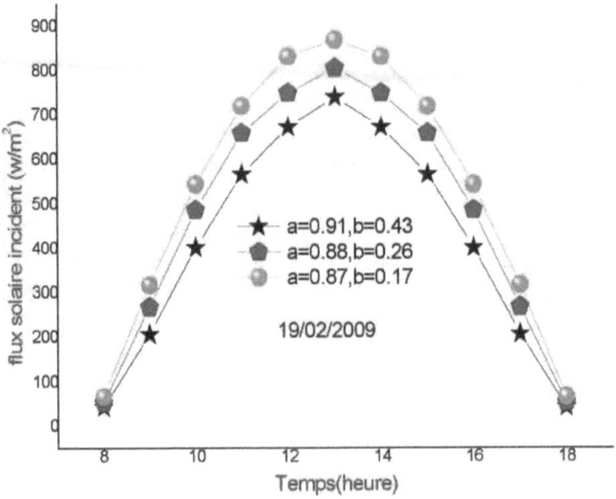

Figure -VI-1-Variation of hourly solar flux according to atmospheric haze

3-Variation in global solar flux

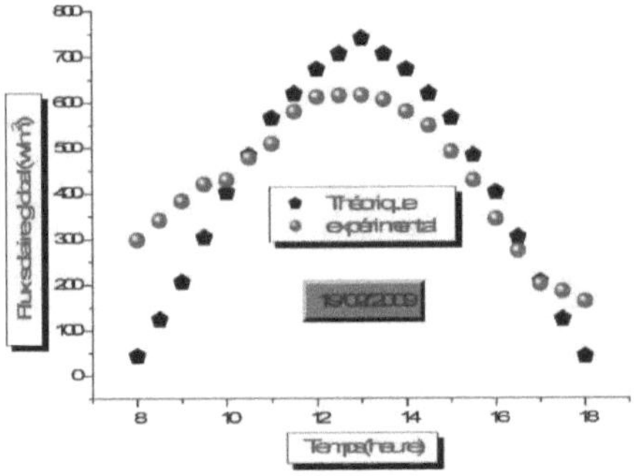

Figure-VI - 2-Variation in global solar flux (city of Ouargla)

The comparison between the measured global solar flux and the one calculated theoretically showed a certain important difference especially from 8 to 10 and from 12 to 15, this difference can be interpreted by the numerical methods used, and also by the materials which we based for the calculation during the realization of the experiment, and also the uncertainties of readers undergone by the user. The state of the atmosphere and the transparency of the clouds during the day of the experiment also have a very important role in the calculation of the global solar flux.

4-The influence of the year

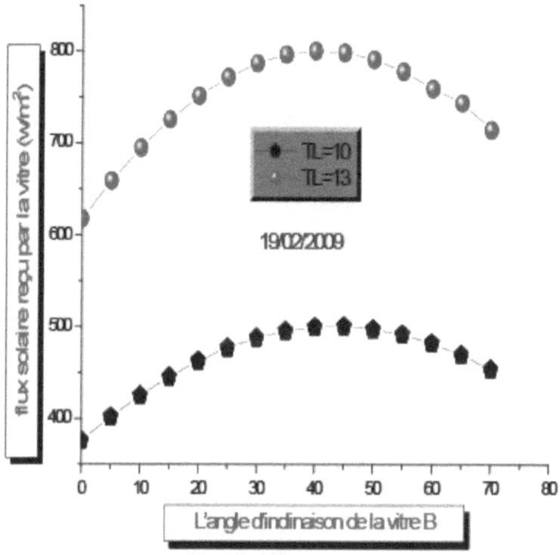

Figure- VI -3- The inf uenc de lan of the íncl naíson on the solaíre flux received by the pane

> Increasing the angle of inclination of the glass leads to an increase in flux. °

The solar flux received by the pane reaches a value of 45, after a decrease in the latter, for this vernal period of the year (19/02/2009 in the city of Ouargla), which can be explained by the decrease in the collection surface. The following figure shows the variation of the solar flux incident on the pane according to the different angles of inclination for two cases of weather.

The best value for the angle of inclination of the window of the solar distiller to receive the maximum incident solar flux is the one with a value close to the latitude of the location under consideration.

5-Time variation in power consumption

800-,

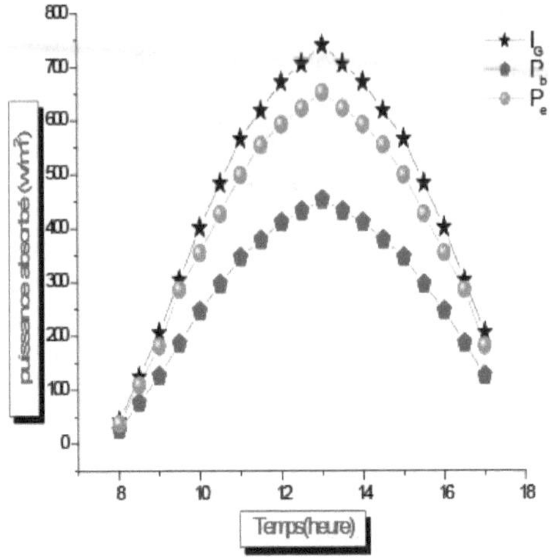

Figure VI -4-Time variation in power consumption.

The powers absorbed by the various components of the solar distiller during the day have a form similar to that of the overall solar flux, the water film absorbs a large quantity of solar radiation which explains their very high overall absorption coefficient which is of the order of 0.882 . The power absorbed by the absorber tank is close to that of the water film, which increases the water temperature. We have noticed a low level of power absorbed by the glass, which can also be explained by the low absorption coefficient of the glass (of the order of 0.05).

6-Variation of external surface temperatures as a function of time

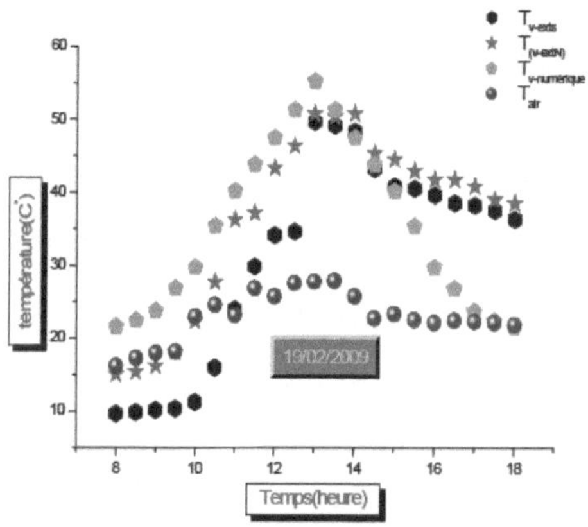

Figure -VI -5-Variation of external surface temperatures as a function of time

The variation in the temperatures of the external faces, either calculated experimentally or estimated theoretically using the calculation code, follows the same pattern as the global solar flux. A certain difference exists between the experimental and theoretical temperature values, which can be explained by the fact that the estimated pane temperature represents the average value of the temperature of the external and internal faces of the pane. The difference in temperature between the two external faces of the pane can be explained by the difference in intensity of the solar flux received by each side of the pane, since it is more intense on the face facing south.

7-Variation of water temperature as a function of time

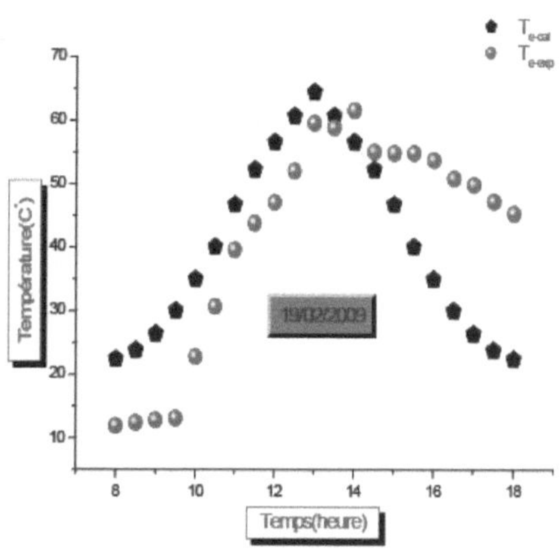

Figure- VI -6-The variation in water temperature as a function of time

> The evaluation of the water temperature as a function of time is either estimated numerically or measured from the experiment having the same appearance as the solar flux, i.e. the shape of a bell. A certain difference between the two estimated and calculated values can be explained by the fact that the power absorbed by the bottom of the distiller is transmitted in its entirety to the water film, but in reality a certain percentage is used to heat

the water in the solar distiller, and also by the influence of the simplifying hypothesis on the criterion: water evaporation is continuous.

8-Variation of the temperature of the absorbent tank as a function of time

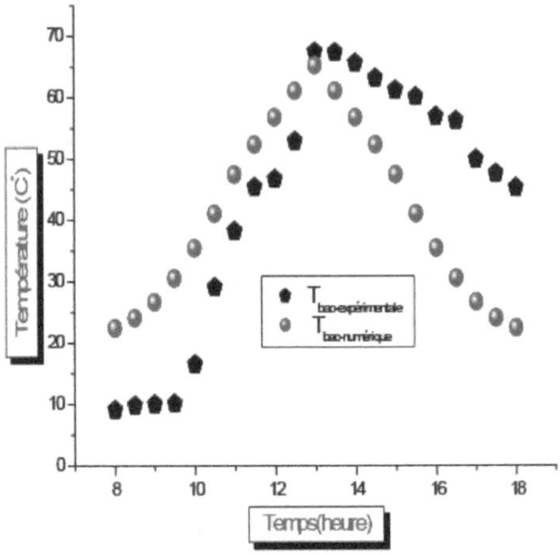

Figure - VI -7-The variation of the temperature of the sorbent tank as a function of time

The variation in the temperatures of the absorber tank, either estimated numerically or calculated experimentally during the day of 19/02/2009, also followed the same shape as the global solar flux, but the graphical representation showed a certain significant difference, especially after 11 o'clock and throughout the time range from midday to 6 o'clock in the evening. This difference between the temperature values can be explained by :

- -The assumption that all the power absorbed by the absorber tray is transmitted to the water film.
- -Cumulative effects in the case of reals, which increase temperatures considerably
- -The coefficient of heat exchange by convection to the outside environment
- -The influence of the thermal properties of materials, which are a function of temperature

9-The effect of wind on the **temperatures of various drive components solar 9-1- 'inf uence on glazing**

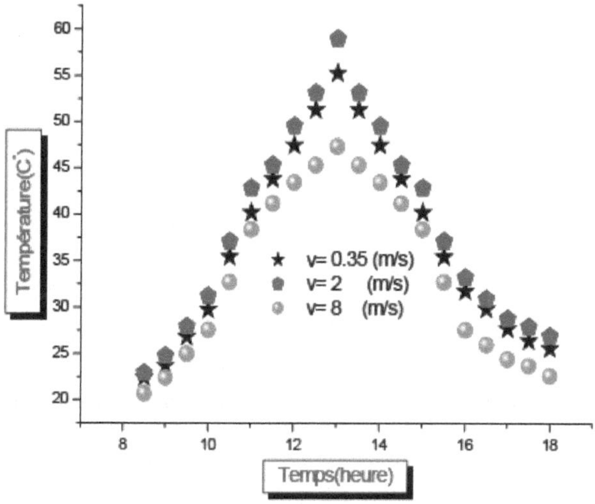

Figure VI -8-Influence of wind speed on window temperature

The graphical representation of the influence of wind speed on the temperature of the pane shows that this parameter has a favourable effect for low wind speeds, which is interpreted by the increase in temperature values, but on the other hand for high wind speed values we have noticed an inverse effect, this unfavourable wind effect can be interpreted by the increase in convection losses to the outside environment.

9-2- 'inf uence le fi m d'eau

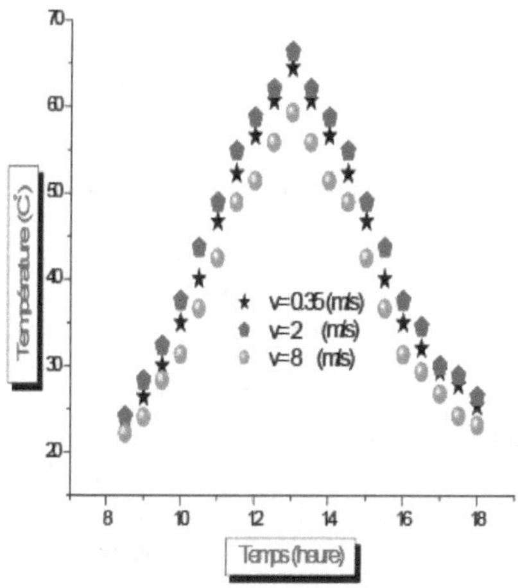

Figure VI -9-Influence of wind speed on water fi lm temperature

The same effect of wind speed on the temperature of the water film inside the solar distiller, but by one degree less, is shown by the slight decrease in the water temperature in the solar distiller for high values of wind speed.

9-3-Influence of wind on 1 absorbent tank

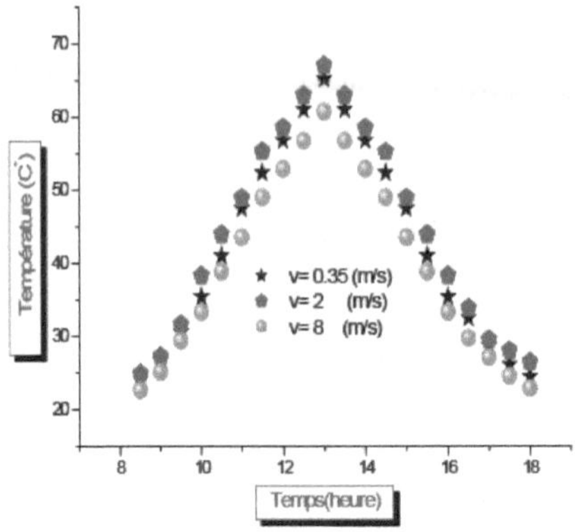

Figure VI -10-The influence of the velocity v nt on the temperature of the absorbent tank

In the case of the absorber tray, we waited for the effect of wind on the glass, but theoretically we found an effect similar to that of wind on the film of water. This phenomenon can be interpreted by the role played by the insulation on the lower side of the absorber tray, i.e. when we plotted the effect of wind on the temperature of the insulation we found the same effect as the glass.

10-Variation of production as a function of time

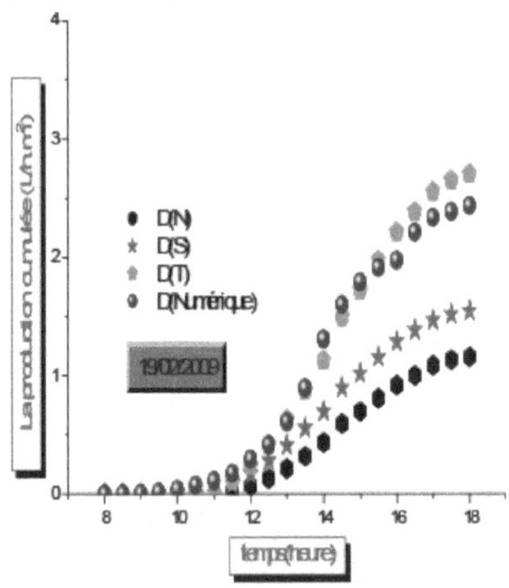

Figure VI -11 -The variation in cumulative production as a function of time

The representation of the variation of the cumulated production, either theoretical or experimental, as a function of time shows the good connection between the theoretically calculated results and those measured during the experiment. A distilled water production value of 2.43 L is found at the end of the day by the calculation code used, but the actual calculation value is 2.7 L, which represents an error. °

of 0.1 /° in estimating the quantity of distilled water produced.

One of the parameters that leads to a difference between the theoretical production and the actual production is the temperature of the pane, which is the average temperature, but water

condensation takes place at the temperature inside the pane.
The representation of the quantity of water produced by the two faces of the distiller, either the one facing south or the one facing north, shows the existence of a certain difference between the two quantities. This difference can be interpreted by the fact that always the face with a low temperature produces a large quantity of distilled water because the drainage process of the water droplets is very rapid on this face.

11-The global cac te as a function of time

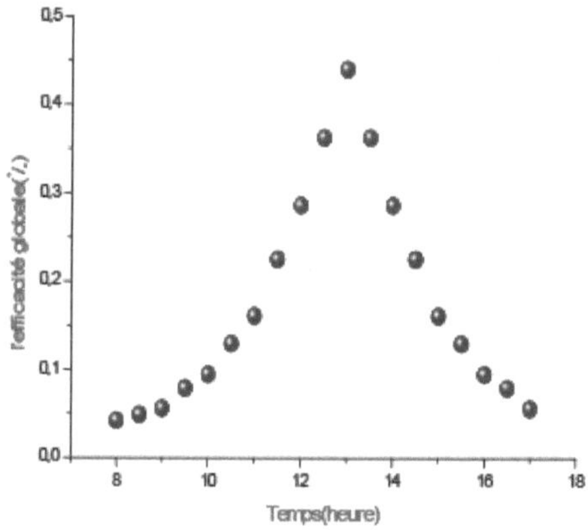

Figure- VI -12-The global water balance as a function of time

The variation of the global efficiency of the solar desalination system also varies in the same way as the solar radiation, the global efficiency of the distiller during the day of the experiment is 44 $^{0}/0$, which really represent the weakness of the output of this Distler water production system, i.e. the following figure which shows the variation of the theoretical global efficiency during the day of the experiment:

12-Variation of the performance factor as a function of time

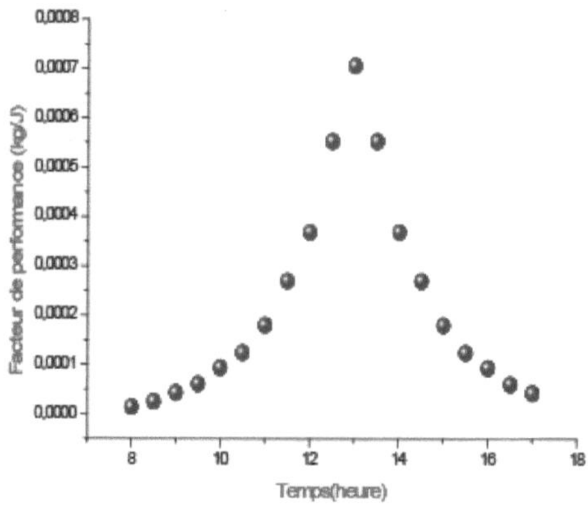

Figure VI -13-Variation of performance factor as a function of time

According to the graph presented in the figure below relating to the variation of the factor of performance of the distiller studied we noticed that the values of this last, are very weak way the maximum value of this characteristic property which reaches at true solar midday is of 7

$.10^{-4}$as the figure shows. The weakness of the values of this characteristic is a response to the poor energy conversions by the solar distiller, which leads to a low production of distilled water.

Conclusion

In this work, we have developed a calculation code to simulate the operating parameters of the conventional solar distiller. Analysis of the results obtained during simulation of the module. allow us to draw the following conclusions:

-The increase in overall solar irradiation, which remains the parameter with the greatest influence on the operation of the solar distiller, leads to an increase in the distiller's operating characteristics.

A higher brine temperature increases production.

Ambient temperature also has an impact on production, as the distiller's low thermal inertia means that it can be operational very quickly. On the other hand, a drop in ambient temperature will lead to a rapid drop in the temperatures of the distiller's components.

The highest value of production is obtained for a value of inclination close to the latitude of the place where the experiment is carried out.

the double-slope solar distiller has the advantage over other types of solar distiller that the north-facing side acts as a collector and the other side acts as a condenser, which speeds up the process of producing distilled water.

There is some difference between the values estimated by the calculation code and the values measured experimentally because of the simplifying assumptions and correlations used to estimate the different heat exchange coefficients for the three heat transfer modes.

The paradox found for the proportionality between the increase in the angle of inclination of the glass and the solar flux received concerns the period when the declination is negative.

General conclusion

In this work we have theoretically studied the double slope solar distiller which operates in direct mode, the latter being based on the greenhouse effect phenomenon. Our interest has focused on the study of the variation of the various operating parameters as a function of the meteorological conditions of the city of Ouargla (solar radiation, ambient temperature, production, efficiencies, etc.).

-From the results found, we have deduced that solar energy is the most important condition for the operation of desalination systems in either direct or indirect mode.

-The temperature values of the water film in the tank are close to those of the absorber tank, which is a little illogical but can be interpreted by the fact that the heat absorbed by the bottom of the distiller is transmitted entirely to the water film, which is justified by the very high value of the water temperature over the whole of true solar noon.

The quantity of water theoretically produced by the solar distiller at the end of the day is 2.43L, but the quantity actually produced is 2.71L, which represents a relative error of $0.1^{0}/0$ in the production calculation. This small production quantity is caused by the high temperature of the glass on the inside, which necessitates the installation of a glass cooling system to speed up the drainage process of the water droplets on the inside of the glass.

Increasing the angle of inclination of the glass leads to a reduction in the solar flux received by the desalination system, which is explained by the reduction in the surface area exposed to solar radiation.

Based on the results obtained and the difficulties affecting the operation and production of the solar distiller, we propose the following lines of research to improve the performance of the distillatemi:

- The use of a self-contained cooling system that can cool the glazing periodically to increase condensation,

integration of the distiller with one of the energy systems (solar collector, wind turbine, etc.)
The use of new low-cost, high-efficiency alloys.

- To develop a standard solar distiller that can be used under any working conditions and in any location.
- Storage of renewable energy to run the distiller during periods when this energy is unavailable, especially in arid and remote regions.

Summary

Water, a vital resource essential to all forms of life, plays a central role in critical sectors such as the pharmaceutical industry, chemical applications and biology laboratories. Faced with the growing scarcity of this resource and the global energy transition, solar distillation is emerging as a sustainable solution, combining economic issues with social challenges. Using an experimental and numerical approach, this work explores the performance of a double-slope solar distiller designed for arid regions. A calculation code developed in FORTRAN language was used to model the key parameters: operating temperatures, overall and internal efficiency, performance factor, and condensate flow rate. The numerical simulations were rigorously compared with experimental data collected under the climatic conditions of the Ouargla region (Algeria), providing a robust validation of the proposed model.

Key words :

Drinking water, Solar distillation, Energy optimisation, Sustainable production, Numerical simulation, Solar energy, Operating temperatures, Arid regions.

References

Abu-Arabi, M. et al. (2002) 'Modeling and Performance Analysis of a solar desalination unit with double-glass cover cooling', Desalination, 143(2), pp. 173-182. doi:10.1016/s0011-9164(02)00238-2.

Abu-Arabi, M. and Zurigat, Y. (2005) 'Year-round comparative study of three types of solar desalination units', Desalination, 172(2), pp. 137-143. doi : 10.1016/j.desal.2004.05.011.

Al-Hinai, H., Al-Nassri, M.S. and Jubran, B.A. (2002) 'Parametric investigation of a double-effect solar still in comparison with a single-effect solar still', Desalination, 150(1), pp. 75-83. doi:10.1016/s0011-9164(02) 00931-1.

Bouchekima, B. et al. (2000) 'Etude Théorique et application pratique du distillateur solaire à film capillaire', International Journal of Thermal Sciences, 39(3), pp. 442-459. doi :10.1016/s1290-0729(00)00216-7.

F. Saida, CH. Bougriou, « Etudes Théorique Et Expérimentale D'un Distillateur Solaire Sous Les Conditions Climatiques de BATNA », *Séminaire International Sur Le Génie Climatique Et l'Energétique, SIGCLE'2010.*

Zurigat, Y.H. and Abu-Arabi, M.K. (2004) 'Modelling and Performance Analysis of a Regenerative Solar Desalination Unit', Applied Thermal Engineering, 24(7), pp. 1061-1072. doi: 10.1016/j.applthermaleng.2003. 11.010.

p] A. A. Sfeir, G.Gharra, « Ingénierie Des Systèmes Solaire Application A L'habitat », *Technique De Documentation* ,1981.

A. Chaker, N. Boukerzaza, "Caractéristiques De Fonctionnement D'un Distillateur Solaire ", *12èmes Journées Internationales De Thermique Tanger, Morocco November 15-17*, 2005.

C. Khelif, B. Touati, "Caractérisation D'un Distillateur A Effet De Serre ", *Rev. Energ. Ren*, 1998.

R. Bernard, G. Meguy, M. Schwartz, "Le Rayonnement Solaire Conversion Thermique Et Applications ", *Technique Et Documentations, Paris*, 256p, 1979.

E. Zayout, L. Bouriden, A. Ahroune, M. Banouni, "Distillation Solaire : Amélioration De La Condensation De La Vapeur D'eau Dans Les Distillateurs Solaires "), *International Forum On Renewable Energies Fire, Tétouan- Maroc Fire, Laboratoire De Thermodynamique Métallurgique Et Rhéologie Des Matériaux, Université Ibn Zohr, Bp28 /S,Agadir,Maroc,* 2002.

http//www.outilssolaire.com

www.col.soufflenheim.ac-strasbourg

B. Nassima, "Etude De L'effet Des Pertes Thermiques Sur Les caractéristiques De Fonctionnement D'un distillateur Solaire ", *magister thesis University of Constantine*, 2009.

B. REMINI "la problématique de l'eau en Algérie du nord ", Larhyss Journal ISSN 1112-3680, N° 08, pp. 27-46, June 2010.
A. Madhlopa, "Development of an advanced passive solar still with separate condenser", pp. 9-12 (2009).
M. Daguenet, "Les séchoirs solaires : théorie et pratique ", Unisco, Paris 1985.
Yves Jannot, "Thermique solaire", p30-p70, October, 2003 Paris (2007).
K. Skeiker, "Correlation of global solar radiation with common geographical and Meteorological parameters for Damascus province, Syria", *Energy Conversion and Management* 47 (2006) 331-345.
M. Yorukoglu and A. Naci Celik, "A critical review on the estimation of daily global solar Radiation from sunshine duration", *Energy Conversion and Management* 47(2006) 2441-2450.
S, Kaplanis and E, Kaplani, "A model to predict expected mean and stochastic hourly global solar radiation I (h; nj) values", *Renewable Energy,* 32(2007)1414-1425.
V. badexu, "Modeling Solar Radiation at the Earth's Surface", pp.1-7(2008).
http:www.rsfa.0fees.net
El-Dessouky, H.T. and Ettouney, H.M. (2002) *Fundamentals of salt water desalination*. Amsterdam: Elsevier.
H. Helene and L. Caroline, "Water consumption in the world" pp.814.
WHO, "Quality standards for drinking water", Third Edition pp.191-197(2004).
O. SOTEHI, A. CHAKER, "Theoretical optimisation of the operation of a capillary film solar distiller", 2009
M. Boukar, A. Harmim and M. Amar, "Construction et expérimentation d'un distillateur solaire vertical indirect à simple effet", *Revue des Energies Renouvelables ICRESD- Tlemcen* 07 (2007) 75 - 81.
Z.A. Maiche, "Desalination of sea water in Algeria", El watan, 30/1/2008.
O.K. Buros, the ABCs of Desalting, P.O. Box241325, Denver, Colorado, 80224, USA 1990.
L. Cherrared, "Amélioration du Rendement d'un Distillateur Solaire à Effet de Serre", *Revue. Energie. Renouvelable, Valorisation* pp 121-124,1999.
F. Livet, "Besoins et Procèdes pour le dessalement de l'eau de SIMP", UMRCNRS 5266, INPG-UJF, mer, BP 75 38402-St Martin d'Heres, France 15 November 2007.
N. Boukerzaza, "Etude des pertes thermiques sur les caractéristiques de Fonctionnement d'un distillateur solaire", *Thesis Magister Université Mentouri de Constantine* 2005.
R. Rahmani, "Rendement d'un distillateur solaire à plusieurs étages, effet de certain paramètre thermo physique sur le rendement", *Thesis Magister Université*

Mentouri de Constantine 2007.

Y. Boualatfi, ‑Investigation de la performance d'un distillateur solaire‑, *Magister Thesis University of Ouargla* 2004.

Tchinda, R. and Kaptouom, E. (2005) 'Simulation Numerique des performances d\\'un distillateur solaire fonctionnant en mode indirect', *African Journal of Science and Technology*, 5(1). doi: 10.4314/ajst.v5i1. 15322.

I. Tabet, ‑Etude d'un distillateur solaire a cascade‑, *Magister thesis, UniversitéMentouri Constantine* 2010.

N. Retiel, F. Abdessemed and M. Bettahar, ‑Etude expérimentale d'un distillateur solaire plan amélioré‑. *Revue des Energies Renouvelables* Vol. 11 N°4 (2008) 635 - 642.

FSC
www.fsc.org
MIX
Papier | Fördert
gute Waldnutzung
FSC® C083411

Kolibri 360 GmbH
Ferdinand-Jühlke-Straße 7
99095 Erfurt, Deutschland
produktsicherheit@kolibri360.de

Druck:
CPI Druckdienstleistungen GmbH
im Auftrag der
Kolibri 360 GmbH
Ferdinand-Jühlke-Str. 7
99095 Erfurt